Measurement of Food Preferences

Measurement of Food Preferences

Edited by

H.J.H. MacFIE
Head of Department of Consumer Sciences
AFRC Institute of Food Research
Reading

and

D.M.H. THOMSON
MMR Product and Concept Research
The Lord Zuckerman Research Centre
Reading

BLACKIE ACADEMIC & PROFESSIONAL
An Imprint of Chapman & Hall
London · Glasgow · Weinheim · New York · Tokyo · Melbourne · Madras

Published by
Blackie Academic and Professional, an imprint of Chapman & Hall,
Wester Cleddens Road, Bishopbriggs, Glasgow G64 2NZ

Chapman & Hall, 2–6 Boundary Row, London SE1 8HN, UK

Blackie Academic & Professional, Wester Cleddens Road, Bishopbriggs, Glasgow G64 2NZ, UK

Chapman & Hall GmbH, Pappelallee 3, 69469 Weinheim, Germany

Chapman & Hall Inc., One Penn Plaza, 41st Floor, New York NY 10119, USA

Chapman & Hall Japan, Thomson Publishing Japan, Hirakawacho Nemoto Building, 6F, 1-7-11 Hirakawa-cho, Chiyoda-ku, Tokyo 102, Japan

DA Book (Aust.) Pty Ltd, 648 Whitehorse Road, Mitcham 3132, Victoria, Australia

Chapman & Hall India, R. Seshadri, 32 Second Main Road, CIT East, Madras 600 035, India

First edition 1994

© 1994 Chapman & Hall

Typeset in 10/12 pt Times by Best-set Typesetter Ltd., Hong Kong
Printed in Great Britain by St Edmundsbury Press, Bury St. Edmunds, Suffolk

ISBN 0 7514 0183 8

A catalogue record for this book is available from the British Library

Library of Congress Catalog Card No: 93-74907

Preface

This book provides comprehensive coverage of the numerous methods used to characterise food preference. It brings together, for the first time, the broad range of methodologies that are brought to bear on food choice and preference.

Preference is not measured in a sensory laboratory using a trained panel – it is measured using consumers by means of product tests in laboratories, central locations, in canteens and at home, by questionnaires and in focus groups. Similarly, food preference is not a direct function of sensory preference – it is determined by a wide range of factors and influences, some competing against each other, some reinforcing each other.

We have aimed to provide a detailed introduction to the measurement of all these aspects, including institutional product development, context effects, variation in language used by consumers, collection and analysis of qualitative data by focus groups, product optimisation, relating preference to sensory perception, accounting for differences in taste sensitivity between consumers, measuring how attitudes and beliefs determine food choice, measuring how food affects mood and mental performance, and how different expectations affect sensory perception.

The emphasis has been to provide practical descriptions of current methods. Three of the ten first-named authors are university academics, the rest are in industry or research institutes. Much of the methodology is quite new, particularly the repertory grid coupled with Generalised Procrustes Analysis, Individualised Difference Testing, Food and Mood Testing, and the Sensory Expectation Models.

Thanks are due to the publishers for encouraging the text in the first place and for preparing the final copy accurately and quickly.

H.J.H.M.
D.M.H.T.

Contributors

A.V. Cardello   Consumer Research Branch, Behavioral Sciences Division/Soldier Science Directorate, Department Troop Support Group, RD&E Center, US Army Natick Research, Development and Engineering Center, MA 01760-5020, USA

M.A. Casey   Minnesota Extension Service, University of Minnesota, 320 Vocational Technical Education Building, 1954 Buford Avenue, St. Paul, Minnesota 55108, USA

M.T. Conner   Department of Psychology, The University of Leeds, Leeds LS2 9JT, UK

S. Edwards   Consumer Sciences Department, AFRC, Institute of Food Research, Earley Gate, Whiteknights Road, Reading RG6 2EF, UK

N. Gains   Reading Scientific Services Limited, The Lord Zuckerman Research Centre, University of Reading, Whiteknights, PO Box 234, Reading RG6 2LA, UK

M.W. Green   Consumer Sciences Department, AFRC, Institute of Food Research, Earley Gate, Whiteknights Road, Reading RG6 2EF, UK

K. Greenhoff   MSTS, 54–62 Station Road East, Oxted, Surrey RH8 0PG, UK

R.A. Krueger   Minnesota Extension Service, University of Minnesota, 320 Vocational Technical Education Building, 1954 Buford Avenue, St. Paul, Minnesota 55108, USA

H.J.H. MacFie   Consumer Sciences Department, AFRC, Institute of Food Research, Earley Gate, Whiteknights Road, Reading RG6 2EF, UK

H.L. Meiselman   Chief, Behavioral Science, Science and Advanced Technology Directorate, Department Troop Support Group, RD&E Center, US Army Natick Research, Development and Engineering Center, MA 01760-5020, USA

H.R. Moskowitz   Moskowitz Jacobs Inc., 14 Madison Avenue, Valhalla, NY 10595, USA

P.J. Rogers   Consumer Sciences Department, AFRC, Institute of Food Research, Earley Gate, Whiteknights Road, Reading RG6 2EF, UK

H.G. Schutz   Division of Textiles and Clothing, University of California, 129 Everson Hall, Davis, CA 95616, USA

R. Shepherd   Consumer Sciences Department, AFRC, Institute of Food Research, Earley Gate, Whiteknights Road, Reading RG6 2EF, UK

P. Sparks   Consumer Sciences Department, AFRC, Institute of Food Research, Earley Gate, Whiteknights Road, Reading RG6 2EF, UK

D.M.H. Thomson   MMR Product and Concept Research, The Lord Zuckerman Research Centre, Whiteknights, PO Box 234, Reading RG6 2LA, UK *now* Wallingford House, High Street, Wallingford, Oxford, OX10 0DB, UK.

Contents

1 A measurement scheme for developing institutional products*

H.L. MEISELMAN

1.1 Introduction

Traditionally, the sensory scientist, consumer researcher, social scientist and nutritionist have had small but significant roles in the design, development and testing of food products. Consumer scientists and market researchers sought to identify or establish a need for products. Sensory scientists studied the sensory properties of food products under development under programs of product improvement, or for quality control during production. The social scientist considered human impacts of food products, and the nutritionist considered nutritional impacts of food consumed (or not consumed). It has been difficult for each discipline to get its voice heard in food research and development. It has usually been beyond reasonableness not only to include each of these but to integrate their roles.

The problems of such integration are perhaps more difficult for the institutional product developer or caterer. Products developed for catering not only consider one product at a time but also consider how to combine products into meals and meals into diets. Those interested in providing food products to hospitals, schools, the elderly, in-plant feeding, specialized remote feeding (e.g. oil rigs), military feeding, etc. all share these concerns. Even those food product companies designing prepared meals for convenience in the home, and special meal products for the person on a weight loss diet, need to consider more than one item at a time in product development.

The United States Army pioneered the development of techniques for the measurement of food acceptance and preference. The nine-point hedonic scale was developed in the 1940s to become a standard measuring scale worldwide (Peryam and Pilgrim, 1957). The use of magnitude esti-

* A number of people at the US Army Natick RD&E Center contributed to the development of the seven-phase measurement scheme described in this article. The major contributions of Drs Edward Hirsch and Armand Cardello are acknowledged. The author presented portions of this paper at the Society for the Chemical Industry's Sensory Panel Meeting in London in March 1990. Part of this manuscript was prepared while the author was a US Secretary of the Army Research Fellow at the Institute of Food Research, Reading, UK.

mation with food products originated with the US Army laboratory (Moskowitz and Sidel, 1971). Once again the US Army has moved ahead in food research, testing and evaluation with the development of a seven-phase testing sequence. This testing sequence was designed for the specific needs of designing food products for the military, but the approach and methods are widely applicable to product developers and caterers. In addition, many basic measurement issues involved in this sequence are relevant to other situations in which consumers and foods interact.

What has been developed at Natick is a sophisticated, complex series of tests for development of products, meals and diets. Such an increase in proposed testing of food products is warranted by several factors. First, development costs for new food products are high. Testing must be seen as a small part of such development. Second, the failure rate with the consumer of many food products is high; if more testing can reduce failures the added costs of testing are worthwhile. Third, in institutional settings such as the military and hospitals, the amount of food waste can be quite high. Improving products through proper testing can reduce this waste thus saving money. Those in the food evaluation and testing community must educate those who use their services that spending more for testing can save money in the long run and can also achieve better quality and better nutrition for the consumer.

It is worth recounting for the reader how Natick developed the following testing scheme. In the late 1970s Natick was asked to test the standard military ration to determine its acceptability for a period of prolonged use i.e. 30 days or more. The ration had been developed for up to 10 days use only. However, product developers are familiar with their products being asked to perform in situations different from what had been the target set for the product originally. The research personnel had a wide range of opinion about what would happen in a 30-day test. Some thought that the active soldiers would be hungry and would eat. Others thought that the soldiers would tire of the same ration system for 30 days, and would increasingly dislike their food. Still others thought there would be a food mutiny leading to the termination of the test. All of the above were wrong.

What happened is that soldiers continued to rate their food highly but ate less and less as the study continued for 34 days. Rated acceptability started high and did not decline over the 34 days. Food intake, as measured by calories, started moderately low and dropped steadily over the test to about 1/2 of what soldiers in the field are expected to eat. Product developers of the army ration were concerned about their product. Army logisticians were concerned about the need for so much food if so much was wasted and nutritionists were concerned about whether soldiers would eat adequately to maintain a vigorous activity schedule. Clearly, those designing and testing military rations had to

Phase I   Consumer Marketing
Phase II   Individual Item Sensory Testing
Phase III   Consumer Meal Testing – Laboratory
Phase IV   Consumer Meal Testing – Field
Phase V   Prototype Testing
Phase VI   Extended Ration Use Validation
Phase VII   Quality Control Testing

Figure 1.1 The seven-phase standardized food testing sequence developed by the US Army.

approach this problem in a different way. Checking each item for acceptability and for sensory properties in the laboratory was not adequate to ensure high field acceptance. What does 'high field acceptance' really mean? Does it mean that soldiers rate their food highly even if they do not eat it? Does it mean that soldiers eat the food? Yes, but why do soldiers' opinions of the food not coincide with their intake? And what type of testing would produce products with high intake by soldiers in the field?

The solution to this problem was a seven-phase testing sequence (Figure 1.1). This sequence was gradually developed by a team of people familiar with basic measurement principles, food product laboratory and field testing and the needs of the military system. The seven phases will be outlined below. The reader should bear in mind that not every product need go through all seven steps. After presenting each step and one or more critical methodological issues of each step, the use of the overall seven-phase sequence will be reviewed.

1.2 Phase I. Consumer marketing

The key to consumer marketing is matching products to consumer interests and needs. It is easier to develop a product or improve a product than to determine that the product is really needed or desired by potential consumers. The traditional approach in much institutional food research and development at this stage is for informal testing and surveying and the strong opinions of management, or the strong desire of the product developer/technologist to produce or improve a specific product. Our goal has been to tie product development at its outset to the consumer (Figure 1.1), and to adapt sound market research techniques to food research and development (see Churchill, 1991, for basic methods).

1.2.1 Whom to test

Whatever consumer testing is carried out at this stage, it is critical that it be conducted with potential consumers of the product. In general, in-house technical or administrative personnel are not appropriate and may have serious biases or may feel pressure concerning new products. Easily available groups, such as students at the nearby university, probably do not reflect the potential product consumer.

If one thinks that the potential product will appeal to a broader range of consumers, then a broad range should be tested. If the product is to fit a market niche aimed at a clear subgroup, then that subgroup should be tested, perhaps along with others that help target the future market.

In any case, an adequate number of people should be tested. Smaller groups, with sample sizes of ten to twenty-five, might be tested at one site or multiple sites, but any group should be large enough to permit separate statistical treatment.

1.2.2   What to test

Techniques to obtain consumer marketing information are basically the entire field of consumer research, and the interested reader is referred to texts in this field. One difficulty at this stage is that one does not have a product to test. Sometimes a product is being improved or modified and the existing product and its characteristics can serve as a basis for questions to consumers. However, when one is looking for new products not yet experienced by consumers the task of defining the problem and parameters to consumers can be more difficult.

This problem faced the laboratory at Natick in its search for novel breakfast foods suitable for military packaging and long shelf-life. Traditional American breakfast items such as eggs, pancakes and waffles could not be provided under the conditions required. When faced with a similar question, fast food providers in the United States developed breakfast sandwiches combining traditional breakfast ingredients but presented in a new form.

The food technologists at Natick depended more on their own ingenuity of what could be produced and less on potential information from consumers about what would be consumed. This resulted in products being designed and produced for testing without prior consumer consensus. Some products failed badly, not on technical grounds but because the consumers would not accept them, while other products succeeded. This project was technology-driven and not consumer-driven.

1.2.3   How to test

There are a number of methods available to collect consumer data (Meiselman, 1988; Churchill, 1991). In general the methods can be divided into questionnaires (written) and interviews (spoken). Either can be structured or unstructured (Table 1.1).

Table 1.1 Methods for obtaining consumer data

	Structured	Unstructured

Questionnaire	Multiple choice items
	Food preference surveys
	Food frequency checklist (moderate
	  to large sample)
		Open-ended
		Essay or questionnaires
		  (small sample)

Interview	Standard questions in the same order
	  (small to large smaple)
		Varying questionnaires
		Focus groups (small
		  sample)

In order to learn about the consumer, the novice or lay person would be tempted to talk to a group of consumers in an unstructured, open-ended way. There are great dangers however in the interviewer biasing the interview and in small numbers of interviewees affecting the conclusion.

Consumer research should be and can be as quantitative and objective as any other type of research. Adequate numbers of respondents should be used for proper statistical analysis. Questionnaires and interviews should be prepared and pilot-tested to ensure consumers understand the questions. Although there is a need for more rigor in consumer research there is a place for more informal, less structured interviews and questionnaires. Asking small numbers of people their opinions can help a consumer researcher to define the issues to be measured in follow-up structured interviews and questionnaires. In fact, it would be foolhardy to design questionnaire and interview formats without first talking to consumers.

The difficulty in consumer research is not in measuring opinion, it is in interpreting that opinion into a course of action. If people like a product it does not necessarily follow they will buy it or use it. If people do not like a product it is not certain they could not be persuaded to try it through advertising. Consumer research provides descriptive data about people and products, attitudes and opinions, reported habits, etc. It is up to the consumer researcher working with the product developers to decide on a course of action.

1.3   Phase II. Individual item sensory testing

Phase II of the Standardized Food Testing sequence comprises traditional sensory evaluation and acceptance testing. The goal is to identify and quantify specific sensory dimensions and characteristics. An additional goal is to relate these to product acceptance.

A great deal of information has been written on individual item sensory testing, including books, reviews, and individual papers (Piggott, 1988;

Meilgaard *et al.*, 1987). Two key issues in sensory testing are worth brief review and discussion: what population to use in actual tests, and what rating scale to use.

1.3.1 Trained and consumer panels

The two different aspects of item testing, sensory analysis and acceptance analysis, typically involve two different types of human testers. Sensory analysis is done with both trained panels and with untrained consumers while acceptance testing is done with untrained consumers only.

Lawless (1984) has added further distinctions to what he calls 'experts' in sensory evaluation. He distinguishes (1) trained panelists with a 'uniform and directed program of training', (2) expert sensory evaluators based on 'long-standing experience with a product' and (3) product developers.

There seems to be general agreement that acceptance testing using hedonic and other affective scales should be done with untrained consumers only. However, any untrained consumer is not adequate. As with Phase I consumer marketing, the consumer tested at Phase II should reflect the target consumer for the product involved. This is probably not often done because of practical constraints of obtaining and maintaining a consumer panel. Furthermore the effect of using groups different from the target groups has rarely been researched and reported.

Shepherd *et al.* (1988) found that ratings of tomato soup samples were different for trained and consumer panels. Only the untrained panel showed a difference in the measure of overall liking. Ratings on a relative-to-ideal scale also differed. In general, the trained and consumer panels preferred different samples. In another study by Shepherd and Griffiths (1987) using the relative-to-ideal scale, based on egg color and flavor, trained and consumer panels agreed on some ratings and disagreed on others.

Lawless (1984) compared experienced and inexperienced wine consumers in describing and matching descriptions to six unlabelled wines. Experienced people matched more wines correctly. Lawless cautioned that these tasters were experienced but were not specifically trained for sensory methods or for wines.

McDaniel and Sawyer (1981) compared a consumer laboratory panel with a consumer house panel in a study of whiskey sour drinks. The 'consumer laboratory panel' was composed of '. . . anyone who liked and was an occasional consumer of whiskey sours . . . The majority of panelists were either students or staff members from the Dept. of Food Science & Nutrition' (p. 182). The consumer home panel also liked and occasionally consumed whiskey sours. Both panels used the 9-point hedonic scale and the magnitude estimation scale. More significant differences in preference were expressed by the home panel. Magnitude estimation resulted in more significant differences in preference in both panels. McDaniel and Sawyer attributed the relative success of the home panel to better motivation, more experience, and the more realistic nature of the test setting. This is a point to which we will return.

Moskowitz *et al.* (1979) tried to model mathematically the relationship between expert and consumer panels in a study of rye bread texture. The expert panel had received formal training in texture profiling and had been using the training for at least one year. The 'consumer' panel was part of the regular taste test panel at the laboratory. The researchers found differences between expert and consumer panels in the criteria or weighting factors used in making sensory ratings of texture. Although similar nonlinear models fitted both the expert and consumer data relating liking of the product to either ingredient levels or physical measures of texture, the values of the parameters in those models were different. The authors claimed that 'expert panel data can be used to predict consumer panel data' (p. 87) and showed equations relating consumer and expert perceptions of texture attributes. In both cases, results were mixed with some attributes yielding better prediction than others.

One overall question which remains from these and related articles is what is meant by experts and consumers. When one strictly defines experts as having received formal training plus experience, and defines consumers as 'the man in the street' with no special training or experience, then very few studies have examined expert–consumer differences.

1.3.2 Choice of rating scales

Sensory testing of individual items in the sensory laboratory involves selecting a proper rating scale for panelists to use. During the 1960s–1980s there was much discussion about which scale was better or best. However, a newer perspective on scaling methodology is developing.

The most frequently used rating scales are the following:

1. The category scale, using a fixed number of possible responses. Either every category is described verbally or quantitatively or only the end anchors and/or midpoint are set (Table 1.2). The best known category scale is the 9-point hedonic scale developed by the US Army in the 1940s and adapted to sensory dimensions as well.
2. The line scale is usually anchored at both ends and the midpoint, but is otherwise not broken into separate defined intervals or categories. The respondent usually marks the line at the place corresponding to the rating.
3. The relative to ideal scale combines the hedonic dimension with the

Table 1.2 Examples of hedonic scales

1. Hedonic category scales
   (a) Jones *et al.* (1955), 9 categories

dislike extremely	dislike very much	dislike moderately	dislike slightly	neither like nor dislike	like slightly	. . .	like extremely

   (b) Pangborn *et al.* (1989), 100 points

0 Dislike	50 Neither like	Like 100
extremely	nor dislike	extremely

2. Hedonic line scales
   (a) Shepherd *et al.* (1988), 100 mm
       Pangborn *et al.* (1989)

Dislike	Neither like	Like
extremely	nor dislike	extremely

   (b) Giovanni and Pangborn (1983), 100 mm

Dislike	Like
extremely	extremely

   (c) Stone *et al.* (1974)

Dislike moderately	Neither	Like moderately

3. Hedonic relative to ideal scales
   (a) McBride (1982), 150 mm

Not nearly sweet enough	Just right	Much too sweet

   (b) Shepherd *et al.* (1988), 100 mm

Not nearly strong enough	Just right	Much too strong

   (c) Shepherd *et al.* (1989), 100 mm

Not nearly salty enough	Just right	Much too salty

4. Hedonic magnitude estimation scales
   (a) Moskowitz and Sidel (1971), first sample served as standard, no zeros or negative numbers.
   (b) Pearce *et al.* (1986), unipolar scale from zero to infinity (liking).
   (c) Pearce *et al.* (1986), bipolar scale from minus infinity (disliking) through zero (neither like nor dislike) to infinity (liking).

intensity dimension to yield a scale going from 'not strong enough' to 'much too strong' with the midpoint labeled 'just right'.

4. The magnitude estimation scale requires respondents to give ratio estimates of strength. For example, if one sample is perceived as twice as strong as another it is assigned a number twice as large.

Details of how to use all these scales can be found in a number of references (for example, Meilgaard *et al.*, 1987).

A number of studies have compared the different rating scales and have assessed users' opinions of the scales. Lawless and Malone (1986a) compared four scales for their efficiency to discriminate among six non-food sample systems including four sandpapers of different grit size. The authors found 'In general, all scaling methods were able to distinguish among the different intensity levels of each product system' (p. 90). When the authors further looked at the size of the F-ratios as an index of discriminating ability, the percent of variance accounted for, and a comparison of *t* values of adjacent stimulus values, all analyses showed a slight superiority of category scaling.

Lawless and Malone (1986b) repeated the study with different consumers and for stimuli that were more difficult to discriminate. Three scales (category, line, line–category hybrid) performed equally well, while magnitude estimation performed as well for the college population only.

Pearce *et al.* (1986) reported a collaborative study coordinated by ASTM involving data from 23 laboratories and involving 553 consumers. Magnitude estimation and category scales were compared in hedonic judgements of eight fabrics. When scale units were converted to similar scales there was strong consistency of results; all fabrics were ordered in the same way, the differences between adjacent samples was similar, and the spread was the same. F-ratios showed similar precision of the three scales.

Giovanni and Pangborn (1983) found that line scaling and magnitude estimation gave similar conclusions about taste intensity and liking of beverages in agreement with the above studies. However, the authors found differences in how the two scales functioned; for example, more restricted ranges were used for magnitude estimation. Pangborn *et al.* (1989) compared category, line scaling and magnitude estimation of caffeine in hot chocolate in a series of five studies. The authors found that category and line scales performed similarly, but they questioned using magnitude estimation for measuring liking.

Lawless and Malone (1986a) and Shand *et al.* (1985) asked consumers their opinions about using different rating scales (Table 1.3). In both studies category scaling and line scaling were ranked better on more criteria and magnitude estimation was ranked worst on most criteria.

At this time, it appears that any established and appropriate rating scale can be used in sensory studies. It is important to select an existing, tested method and not develop a new one without extensive testing. There appears to be a slight advantage for category scaling in both discrimination and user appeal, and a slight disadvantage to magnitude estimation.

Table 1.3 Ranking of three scaling methods for three criteria in two studies of sensory assessment

	Category scaling	Line scaling	Magnitude estimation
Criteria[a]			
Ease of learning	1.4	1.6	3.0
Effort for sample evaluation	1.6	1.6	2.8
Scale preference	1.4	2.1	2.6
Criteria[b]			
Easiest to understand	1.7	1.5	2.7
Fastest to complete	1.5	1.5	3.0
Least restrictive	2.5	1.0	2.5

[a] *Source*: Shand *et al.*, 1985.
[b] *Source*: Lawless and Malone, 1986b. Rankings were averaged for the ratings of best and worst methods.

1.4   Phase III. Consumer meal testing – laboratory

The previous phases on consumer research and item sensory testing both involve established research and testing areas. Each has a large literature of methods, theory and case histories. The next two phases of standardized testing (Figure 1.1), Phase III: Consumer meal testing – laboratory and Phase IV: Meal testing – field, have far less background. This is because both testing and research in foods have tended to focus on either items or diets, the latter from a nutritional perspective.

1.4.1   What is a meal?

Combinations of foods or meals have been overlooked for a number of reasons. First, the line between what is a meal and what is an item is gray or blurred at best, and non-existent at worst. A dictionary definition of a meal is 'an eating experience' which opens the field to any item or combination. Traditional views would emphasize traditional meal courses (appetizer or starter, salad, vegetable, potato, meat, fish or casserole, dessert, beverage). Lately, the context of the meal is receiving increased attention, and the presence or absence of social and contextual aspects could actually define a meal.

Even when one is comfortable that one is studying meals rather than items, the techniques can become rather complicated. There are no established procedures for studying meal acceptability. Portion sizes, proper heating, proximity on the plate, etc. all become issues. Should the consumer experimental subject judge each item separately? Must the meals be served at proper meal times?

The evaluation of meals has without doubt increased as the food product industry and food service or catering industries have gone more heavily into prepared meals or ready meals. Most or all of these data are proprietary and not available to the researcher or product tester.

1.4.2 Acceptance and consumption

For Phases III–VI, the criterion used to assess a product (a meal or diet) is a combination of acceptance and consumption criteria. This is an important point which should not be overlooked. It is not enough to select one measure, acceptance or consumption. We have learned that these measures mean different things and that both measures are needed for complete information. Farleigh et al. (1990) reviewed the relationship between total salt intake and rated acceptance and found very mixed results from a number of studies. Even when the review was narrowed to acceptance and table salt use alone, the relationship between rated acceptance and intake was not clear. The authors note the large number of factors determining salt intake as well as other food choices. It is worth noting that even for a relatively small isolated part of overall rating, i.e. salt intake, one cannot specify the acceptance–intake relationship.

Obviously how much one weights the acceptance data and the consumption data will vary with the type of test. It appears to the present author that acceptance data are weighted more heavily in the laboratory and under less realistic field tests, while consumption data should be weighted more heavily in more realistic field tests.

Acceptance measures have been dealt with above, and as noted, in a number of texts and articles. Consumption measures have also been widely discussed but with even more confusion (Bingham, 1987). The following range of measures is available:

- selection rates for items;
- selection rates for individuals;
- aggregated food intake;
- individual food intake.

The simplest measure is probably just determining how many people take each item. If one starts with a set number of servings, then this is easily determined if second servings are not permitted. These data of course give no information on meal combinations or individual behavior. To determine those, one needs to know the selection rate for individuals, i.e. what items were picked by each person. Automated cash registers with keys for each item on the menu can give either of these measures depending on whether each person's selections are stored separately or aggregated (into total hamburgers, total French fries, etc.).

Generally, one does not only want to know what was taken but what was actually consumed. This involves intake methodology. The simplest approach using direct weighing rather than recall or record procedures is aggregated weighing. If 100 people eat 40 pounds of beef, and 20 pounds of carrots, etc., then the average consumption per person can be calculated. Such data do not show the distribution of individual intakes and do not permit statistical analysis, but do give a quick reference figure.

Individual weighed intake is the gold standard measure of nutritional intake. Each food for each person is measured to give a total dietary intake. This approach can be labor intensive and slow for large numbers of testers and for large numbers of foods.

1.5   Phase IV. Consumer meal testing – field

Phase IV, the testing of meals under more realistic, non-laboratory conditions, represents the first phase of the 7-phase sequence which is conducted in the field (Figure 1.1). There are a number of important considerations involved in moving the research setting from the laboratory to the field.

1.5.1   Realism

Obviously the laboratory is not a natural eating environment for anyone, but not all eating situations are 'natural' either. Is the test site where the people normally eat? Is this instance typical of other field environments for these people? Is the meal being evaluated being designed for situations like this? Ideally, food designed for the home should be tested in the home, and food designed for hospital use should be tested in the hospital, etc. But this is not always done. Factors of cost, convenience, tester and test site availability play a big part. Testing in the field is not an automatic guarantee of realism; one must test in the environment where the product will be used. Environment here includes not only physical characteristics but social, economic, and other characteristics as well.

1.5.2   Test population

Getting enough of the right kind of people for enough time is the goal. One usually has to recruit more than needed to account for subject attrition. Longer studies might require some financial or other incentive to remain in the study. And what kind of people participate can influence the outcome. One needs to test the meals on a sample of the population for whom the meals were designed. This is usually difficult to achieve. Volunteers do not always conform to population distributions; personnel assigned to the study (non-volunteers) can pose problems of representativeness and of motivation.

1.5.3   Adhering to test protocol

Studies done in both the laboratory and the field usually have test proto-cols or plans detailing how the study will be run. Factors such as test duration, group sizes, supervision, data collection, etc., are all detailed. The advantage of laboratory research is that usually one can follow the test protocol closely. The reality of field research is that deviations from test plans are often necessitated by factors outside the control of the researcher.

1.5.4   Item and meal acceptance

The relationship between overall meal acceptance and the acceptance of individual meal components is not well understood. Although there have been a few papers modelling the relationship between meal scores and item scores, the form of the mathematical equation and other aspects remain unclear.

Investigators have made a number of attempts to describe and predict how individual meal items or menu items combine to form a meal. Eindhoven and Pilgrim (1959) were perhaps the first to approach this quantitatively. They argued that overall meal acceptability is not the sum or average of individual meal components, i.e. that preferences for single foods are non-additive. Additive models proposed by others, years later, will be described below.

Eindhoven and Pilgrim obtained ratings on 9-point category scales ranging from 'like extremely' to 'dislike extremely' and from 'go with extremely well' to 'go with extremely poor'. Ratings were obtained for 57 single food items and combinations of those items. The food items were 22 main dishes, seven potatoes, three other starches, ten vegetables, five soups, five desserts, and five salads. Various two-item combinations (main dish with another item) were tested; not all possible combinations were tested. Interaction between main dishes and side dishes was demon-strated in two ways: by significant interaction in the ratings of the two item combinations, or by significant correlation of combination. In the former case, all interactions except desserts and salads were significant on the compatible scale but only potatoes, starches, and some vegetables were significant on the hedonic scale. Similarly, correlations of less than 0.75 were obtained for all items on the compatible scale except salads, but only for potatoes and some vegetables on the hedonic scale. Cor-relations on the compatible scale were much lower in every case than for the hedonic scale, indicating more item interaction. Overall, potatoes and other starches showed greatest interaction and vegetables next greatest. Soups, desserts and salads showed little interaction with main dishes; therefore this combination with main dishes can be predicted well from the preferences for the individual foods.

Rogozenski and Moskowitz (1982) obtained overall meal ratings from 173 respondents using the 9-point hedonic scale. Meals consisted of 136 different combinations of main dish ($n = 70$), starch ($n = 16$), vegetable ($n = 26$), salad ($n = 13$), and dessert ($n = 16$). Ratings were obtained for individual items as well as ten 5-item combinations. Multiple linear regression was used to obtain values for a linear model of meal acceptability. The authors found that meal preference could be predicted to within 0.35 scale points for averaged data, and that the linear model accounted for 71% of the variance.

Rogozenski and Moskowitz calculated the normalized weights of their five meal components as follows: main dish 49%, starch 16%, vegetable 12%, salad 7% and dessert 16%. It was noted, however, that there were large individual differences in meal component weights for all meal components. The widest range of weights was for main dishes. These data demonstrate the difficulty of applying menu models to small population groups, to population subgroups on whom normative data are not available, and to foods for which normative data with the same population are not available.

Mean acceptance ratings for the 136 meals almost all fell between 5.0 and 7.0, with one sardine meal at 4.20 and two beef meals (hamburger, grilled steak) at 7.10. It is interesting to question whether such a narrow range of meal scores results from the items surveyed or the method employed, or reflects the genuine narrow acceptance range of real meals.

Turner and Collison (1988) obtained acceptance scores just after consumption of meals composed of starter, main dish, potato, vegetable, and sweet. Scores were obtained for overall meals and for individual items for nine meals. Using multiple regression to fit an additive model of meal acceptance, they obtained the following:

$$\text{whole meal} = 0.57 + 0.43 \text{ (main dish)} + 0.21 \text{ (sweet)}$$
$$+ 0.14 \text{ (starter)} + 0.14 \text{ (potato)}$$

The correlation coefficients for the whole meal and for each meal component ranged from 0.41 for meal and starter to 0.71 for meal and main dish. This supports the importance of the main dish in contributing to overall meal acceptability. Correlations of other meal components with each other ranged from 0.22 to 0.48 and were all statistically significant. The authors attributed this to the tendency of respondents to rate foods consistently high or low. However, ratings for all 285 meals averaged 7.1 and component averages ranged from potato (6.8) to sweet (7.3). This is a very narrow range of scores for such items as the authors point out. The effect of such a narrow range on correlations, differences, and models is unknown.

Turner and Collison made some interesting observations which could serve as the bases of further study: younger customers (aged < 26) scored lower, catering students or staff scored lower, regular diners scored higher, and certain tables of diners differed from other tables. This latter effect could reflect social influence as recently demonstrated by Engell *et al.* (1990).

Preference ratings from 22 customers obtained at least one week prior to actual eating were compared with actual item acceptance values obtained at the meal. Acceptance was higher than preference for 34/36 items leading the authors to suggest that such surveys might underestimate the acceptance of good quality food. However, it is usually assumed that preference surveys reflect the attitude toward highly preferred food (Schutz and Kamenetzky, 1958).

Rogozenski and Moskowitz (1982) also attempted to model how preference for food changes since time of last survey. They obtained 9-point hedonic scale ratings for 144 foods. Respondents were asked to imagine they had last been served the food three months ago, one month, two weeks, one week, three days and yesterday. Data were obtained from 251 respondents but only 173 surveys were deemed usable. The high percentage of unusable surveys indicates either the difficulty of the task or the resistance of the subjects to the task. The authors suggested that the time preference relationship is best described by a logarithmic quadratic function.

More research is needed to understand how meal components contribute to overall meal acceptability. The complexity of this problem is overwhelming because one must consider at least the following: ratings of food items in isolation, ratings of food items in various combinations, and ratings of food items and combinations dependent on time of last serving. Many studies have utilized paper and pencil approaches. Perhaps a great deal can be learned from collecting acceptance data from real eating environments and modelling meal acceptability.

1.6   Phase V. Prototype testing

The first four phases of standardized food testing involved short-term tests in the laboratory or in the field. We now turn to the more extensive field tests required for products designed for longer term use (Figure 1.1). Such tests are required when a series of meals has been designed and its impact on consumers must be tested. Examples of Phase V tests in institutions include the following:

- Testing a new or replacement military ration designed for one week use.
- Testing a new ten-day ready meal system for hospitals.
- Testing a two-week meal cycle for oil rig workers.

- Comparing fast food versus conventional food in a college cafeteria for one week.

As we shift the focus from items and meals to diet, there is a shift in attention toward nutritional, physiological and health factors. Acceptance of meal items (and consumer satisfaction with the overall diet) is still measured, but the main focus shifts to what the consumer eats and its possible impact on health and well-being. The acceptance data help to explain the consumption pattern and help to form the basis for future item changes.

For the first time also, in Phases V and VI, the food studies are designed as comparisons between two conditions, one serving as a control group and one serving as an experimental group. This approach was determined after much deliberation and experimenting with different criteria for determining diet success or failure. It was concluded that more meaningful data were collected in a comparative mode. The danger in testing just the new ration is that it provides no means of saying how conditions of the test – remember this is a realistic field test – might have skewed the data. Very high or low intakes, or very high or low acceptance scores are more easily interpreted when compared to a baseline.

Probably the largest collection of publicly available test reports on prototype testing are the food studies done by the US Army Natick Research Development and Engineering Center, the developer of US military rations and by the US Army Institute of Environmental Medicine, the persons responsible for medical and nutritional aspects of US Army rations. Both organizations are located in Natick, Massachusetts. Reviews of some of these studies have been published in several sources including the article by Meiselman *et al.* (1988) and the more recent review by Hirsch and Kramer (1991). In addition, individual reports on each study are generally available for those interested in detailed results.

A number of US Army studies have focused on the main military ration, the Meal Ready-to-Eat (MRE) (Table 1.4). This meal is composed of various meal components each packed in a flexibly packaged pouch. There are twelve menus available. Prototype testing can be relatively simple using visual estimation or self report of food intake plus food acceptance ratings or it can involve more measures (for example see Table 1.5, from Popper *et al.*, 1987). The requirements of the project and the decisions to be made based on data will determine the level of data collection. The study by Popper *et al.* (1987) provided the key data leading to adoption of a new ration involving expenditures of tens of millions of dollars.

The two main types of human data relevant to the current discussion from such studies, food intake and food acceptance, are not necessarily highly correlated. As Meiselman *et al.* (1988) pointed out, food intake

Table 1.4 Prototype testing of Meal Ready-to-Eat ration

Reference	Specific ration	Test duration (Days)*	Caloric intake	Acceptance	
				Overall	Main dish
Askew et al. (1986)	MRE IV	12	2282		NA
Popper et al. (1987)	MRE IV	11	2517		5.7
	MRE VII		2517		6.8
	IMP MRE		2842	7.86	7.6
Engell et al. (1987)	MRE V	10	2733		6.2
		(4 MREs/day)			
Morgan et al. (1988)	MRE VIII	11	3217	–	–
		(4 MREs/day)			
Lester et al. (1989)	MRE VIII	3	2550		7.03
		(4 MREs/day)	2289		7.39
			2206		6.56
Edwards et al. (1989)	MRE VI	10	2009		6.34
	MRE VIII	(4 MREs/day)	2802		7.33

* 3 MREs per day unless noted.
IMP = improved. NA = not available.

Table 1.5 Data collection schedule for 11-day study

	Base-line	Day 1	Day 2	Day 3	Day 4	Day 5	Day 6	Day 7	Day 8
Body weight	X	X	X	X			X	X	
Urine specific gravity	X	X	X	X			X	X	
Food/water intake		X	X	X			X	X	
Food acceptance				X				X	
Final questionnaire									

Source: Popper et al., 1987, Table 2.

data appear to track improvements in the ration much better than acceptance data. The latter are affected by a broad range of social, situational, and economic factors which are just beginning to be identified (Hirsch and Kramer, 1991). For those interested in measuring acceptability in the laboratory it is a challenge to understand what acceptability means in the field. It appears that food intake is a more comprehensive measure of all the factors influencing the consumer than is food acceptability. Table 1.4 lists both caloric intake and the rated acceptance of the main dishes for six prototype studies of the Meal Ready-to-Eat.

When comparing acceptability scores from different studies one must consider how the data were collected. Acceptance data collected at the mealtime are consistently higher than acceptance ratings collected in a post-test questionnaire (Popper et al., 1987; Edwards et al., 1989). Recall that Turner and Collison (1988) similarly found higher ratings at meal time than in a pre-test questionnaire. This is possibly because at mealtime only those choosing to eat the item are rating it, whereas in a post-test questionnaire those who would not have selected the item to eat or who would have traded it are also rating it. Or it could reflect other measurement differences.

1.7   Phase VI. Extended ration use validation

This phase of field testing is appropriate for situations in which a food system is designed for extended use (Figure 1.1). In military situations we have somewhat arbitrarily set a duration of over 30 days as being extended use. For other purposes, different durations could be used. Extended testing would be appropriate for the following situations:

- Comparing students who ate a traditional lunch with students who ate a diet drink as their sole lunch.
- Comparing different meal products for long-term hospital patients; for example, special milkshakes versus blended regular food for jaw surgery patients.
- Comparing different special rations on two groups of daily long distance runners.
- Studies of food monotony in which the same diet is served repeatedly.

Whereas in Phase V, prototype testing, the focus shifted to food intake and to comparative testing, in Phase VI the focus shifts even further to include a number of health measures. It is essential in tests such as those used as examples above to ensure that the health of the participants is not being negatively affected. Nutrition status and hydration status show whether basic food and liquid intake are in balance or deficit. Measures of health disorders can be assessed using a standardized symptoms checklist. As with other measurements, developing one's own checklist can be very time consuming and can require an entire separate project. It should only be attempted when no existing form can be used, and then, adequate time and resources must be made available to design and pretest the new form. The US Army Institute of Environmental Medicine in Natick, Massachusetts, has developed several symptoms questionnaires (Sampson and Kobrick, 1980). The currently used version is shown in Figure 1.2. Another version was used by Hirsch et al. (1984), Askew et al. (1987b) and others (Table 1.6).

In addition to getting information on whether experimental subjects in long term studies are experiencing adverse health systems, some studies try to directly measure aspects of physical and mental performance which might be compromised (Tables 1.7 and 1.8).

As with prototype testing in Phase V, a number of Extended Use tests

Printed in U.S.A.   NCS Trans-Optic® MP30-35481-321   A6000

NAME: _____   TIME (24 hour): _____

ENVIRONMENTAL SYMPTOMS QUESTIONNAIRE

IMPORTANT DIRECTIONS FOR MARKING ANSWERS	• Do not use ink or ballpoint pen. • Make marks that fill the circle completely. • Erase cleanly any answer you wish to change. • Do not make any stray marks on this form.	EXAMPLES: CORRECT MARK ● INCORRECT MARKS ⊗ ◔ ◔ ⊙ ✺
USE NO. 2 PENCIL ONLY		

Fill in the response for each item to correspond to HOW YOU HAVE BEEN FEELING
DURING THE PAST DAY/NIGHT. Please answer EVERY item. If you did not experience
the symptom fill in the response labeled "NOT AT ALL".

DESCRIPTION OF SYMPTOM	NOT AT ALL	SLIGHT	SOME WHAT	MODERATE	QUITE A BIT	EXTREME
1. I FELT LIGHTHEADED	0	1	2	3	4	5
2. I HAD A HEADACHE	0	1	2	3	4	5
3. I FELT SINUS PRESSURE	0	1	2	3	4	5
4. I FELT DIZZY	0	1	2	3	4	5
5. I FELT FAINT	0	1	2	3	4	5
6. MY VISION WAS DIM	0	1	2	3	4	5
7. MY COORDINATION WAS OFF	0	1	2	3	4	5
8. I WAS SHORT OF BREATH	0	1	2	3	4	5
9. IT WAS HARD TO BREATHE	0	1	2	3	4	5
10. IT HURT TO BREATHE	0	1	2	3	4	5
11. MY HEART WAS BEATING FAST	0	1	2	3	4	5
12. MY HEART WAS POUNDING	0	1	2	3	4	5
13. I HAD A CHEST PAIN	0	1	2	3	4	5
14. I HAD CHEST PRESSURE	0	1	2	3	4	5
15. MY HANDS WERE SHAKING OR TREMBLING	0	1	2	3	4	5
16. I HAD A MUSCLE CRAMP	0	1	2	3	4	5
17. I HAD STOMACH CRAMPS	0	1	2	3	4	5
18. MY MUSCLES FELT TIGHT OR STIFF	0	1	2	3	4	5
19. I FELT WEAK	0	1	2	3	4	5
20. MY LEGS OR FEET ACHED	0	1	2	3	4	5
21. MY HANDS, ARMS OR SHOULDERS ACHED	0	1	2	3	4	5
22. MY BACK ACHED	0	1	2	3	4	5
23. I HAD A STOMACHACHE	0	1	2	3	4	5

Figure 1.2 Part of the environmental symptoms questionnaire developed by the US Army
Institute of Environmental Medicine.

Table 1.6 Extended testing of Meal Ready-to-Eat ration

Reference	Specific ration	Test duration	Caloric intake	Acceptance Overall	Acceptance Main dish
Hirsch *et al.* (1985)	MRE IV	34	2189	7.05	7.05
Hirsch *et al.** (1985)	MRE IV	45	3149	6.05	
Askew *et al.* (1987b)	MRE VI	30	2782	6.53**	6.99

* Student volunteers.
** Final questionnaire.

Table 1.7 Test points during RLW-30 30-day study

	Body weight	Body fat	Endurance	Strength	Blood	Urine specific gravity
Pre-experiment	x	x	x	x	x	x
Week 1	x			x		x
Week 2	x			x	x	x
Week 3	x			x		x
Post-experiment	x	x	x	x	x	x

	Food intake	Physical exams	Physical symptoms	Human factors	Cognition vigilance mood morale
Pre-experiment		x	x	x	x
Week 1	x		x	x	x
Week 2	x		x	x	x
Week 3	x		x	x	x
Post-experiment		x	x	x	x

Source: Askew *et al.* 1987.

Table 1.8 Testing schedule for prolonged feeding of Meal Ready-to-Eat (MRE) rations

Measures	Frequency	When
1. Food-related measures		
a. Food preference	4×	Baseline, T_1, T_2, T_3
b. Food acceptability	11 days	Periods A, B, C, D
c. Food and water consumption	11 days	Periods A, B, C, D
2. Nutritional status		
a. Body weight	4×	Baseline, T_1, T_2, T_3
Non-volunteers	2×	Baseline, T_3
b. Anthropometry, height, skinfold thickness	2×	Baseline, T_3
c. Body fluid status	4×	Baseline, T_1, T_2, T_3
d. Blood constituents	4×	Baseline, T_1, T_2, T_3
3. Clinical symptoms		
a. Symptoms checklist	4×	Baseline, T_1, T_2, T_3
b. Weekly availability of physician		
4. Psychological tests		
a. Cognitive and psychomotor performance	4×	Baseline, T_1, T_2, T_3
b. Mood	4×	Baseline, T_1, T_2, T_3
c. Morale and perceptions of leadership	4×	Baseline, T_1, T_2, T_3

Source: Hirsch *et al.*, 1984.

have been conducted on the US Army ration, the Meal Ready-to-Eat (MRE). Two of the tests shown in Table 1.6 were conducted with military personnel and one test was conducted with university students. As with prototype testing the results of caloric intake and food acceptance ratings do not track well. That is, people did not necessarily eat more of the food they liked more. In fact, the highest rated ration was consumed (7.05) at the lowest level (2189 kcal/day). The student test however also shows the dramatic differences in food intake which can come about with changes in test setting and test population. When fed the identical military ration, the students ate almost 1000 kcal/day more than military personnel in the field (3149 versus 2189 kcal/day). Meiselman *et al.* (1988) attributed this difference to the effects of the various situational factors rather than to the effects of the food. In a series of studies Hirsch, Kramer, Engell and their colleagues at Natick have identified a number of these situational variables and demonstrated their potent effects on food intake (Hirsch and Kramer, 1991). In parallel studies, Meiselman has shown the same effects of certain situational variables in a civilian feeding establishment in the United Kingdom, and has added the economic variable of meal cost. It cannot be over-emphasized that food developers, food evaluators, and food habits researchers have overly focused on the food itself and have largely ignored the non-food factors that probably have far greater control over food acceptance and consumption.

1.8 Phase VII. Quality control testing

After a food has been successfully developed and put into production, there can be a reduced interest in its effects on consumer behavior. Of course there is great interest when the consumer rejects the product. This is true in institutions where rejected food leads to waste and in food companies where rejected food leads to reduced sales. The purpose of Phase VII, quality control testing, is to avoid some of these crises by periodically testing the food to ensure that the food itself is not changing and leading to consumer rejection (Figure 1.1).

It is easy to conceptualize quality control testing within the context of the seven-phase testing sequence. Recall that Phase II was individual item sensory testing. The task in Phase VII is to ensure that the food product scores equal to or greater than its score at Phase II. The use of reliable quantitative testing methods at both Phases II and VII will make this task easier. Of course it is important to test the products at Phase VII in the same manner with the same product presentation, panel make-up, etc.

1.9   How to use the seven-phase testing sequence

How can the seven-phase testing sequence be applied to a variety of problems in research and development? Some situations will require use of the entire seven-phase series. A prime example is, of course, the development of a military ration intended as the sole source of food for 30 days. Such a ration would require concept testing, item testing, meal testing, and finally field testing for short and then long durations.

However, other problems would require only one or several phases. Let us work through several examples. An airline would like to compare its current frozen ready meal with an alternative containing fewer meal components. Eventually, this needs to be tested at Phase IV, meal testing-field, but work should begin at Phase I, consumer testing. This initial work would obtain consumer reaction to the current system and to the proposed system. Changes might need to be made to the proposed system based on consumer responses. Items from the proposed system would require laboratory sensory testing if no testing had been conducted. This is the quickest, least expensive way of identifying and correcting problem food items. This could be followed by combining items into meals in the laboratory. Finally, the proposed meals would be compared with the existing meals in an airline (field) test. Any problems found at Phases II–IV might require additional work at earlier phases.

For a second example let us consider how to develop a one-week hospital food system. Perhaps the hospital considers it can repeat its menu weekly because the average hospital stay is less than one week. Again, work would begin by surveying people about possible catering concepts. Surveys might include patients and also hospital dietitians and nurses. Let us assume that a concept for catering had been selected and that the items had already been tested in the laboratory as items and meals by a contractor in a previous test. If the data were thought to be applicable for Phases II and III then we could proceed to hospital meal testing (Phase IV) and finally to prototype testing (Phase V). Hopefully the latter test would compare the new system to an existing system.

In the third example, consider a 30-day packaged ration designed for long-distance backpackers. The ration is supposed to contain all nutrients for 30 days and not require supplementation. The packaged ration would consist of totally new items not available previously. After concept testing with consumers (Phase I), the items and meals would require testing at each phase followed by extended use testing (Phase VI), emphasizing nutritional and performance criteria. It might be tempting to skip certain phases to save time and/or money, but the risks should be kept in mind. Speeding up development could result in the need to return to earlier test phases to collect needed information, or even worse, could result in a completed product which fails at Phase VI.

The seven-phase testing sequence presents a carefully thought out series of tests which cover many requirements of food product development and catering system development. By carefully applying the seven phases to each problem, the developer can increase the chances of producing a product which will be acceptable to the consumer.

References

Askew, E.W., Claybaugh, J.R., Cucinell, S.A., Young, A.J., and Szeto, E.G. (1987a) Nutrient intakes and work performance of soldiers during seven days of exercise at 7200 ft altitude consuming the Meal, Ready-to-Eat ration. Technical Report T3-87, US Army Research Institute of Environmental Medicine, Natick, MA.

Askew, E.W., Munro, I., Sharp, M.A., Siegel, S., Popper, R., Rose, M.S., Hoyt, R.W., Martin, J.W., Reynolds, K., Lieberman, H.R., Engell, D.B., and Shaw, C.P. (1987b) Nutritional status and physical and mental performance of special operations soldiers consuming the Ration, Lightweight or the Meal, Ready-to-Eat military field ration during a 30 day field training exercise. Technical Report T7-87, US Army Research Institute of Environmental Medicine, Natick, MA.

Bingham, S.A. (1987) The dietary assessment of individuals: methods, accuracy, new techniques and recommendations. *Nutrition Abstracts and Reviews (Series A)*, **57**, 705–42.

Churchill, G.A. (1991) *Marketing Research Methodological Foundations*, 5th edn, The Dryden Press, Chicago.

Edwards, J.S.A., Roberts, D.E., Morgan, T.E., and Lester, L.S. (1989) An evaluation of the nutritional intake and acceptability of the Meal, Ready-to-Eat consumed with and without a Supplemental Pack in a cold environment. Technical Report T18-89, US Army Research Institute of Environmental Medicine, Natick, MA.

Eindhoven, J. and Peryam, D.R. (1959) Compatibility of menu items. *Quartermaster and Container Inst.*, Report 35–9.

Engell, D.B., Kramer, F.M., Luther, S., and Adams, S.O. (1990) The effect of social influences on food intake. *Society for Nutrition Education*, Anaheim, California.

Engell, D.B., Roberts, D.E., Askew, E.W., Rose, M.S., Buchbinder, J., and Sharpe, M.A. (1987) Evaluation of the Ration, Cold Weather during a 10 day cold weather field training exercise. NATICK Technical Report TR-87/030, US Army Natick Research, Development and Engineering Center, Natick, MA.

Farleigh, C.A., Shepherd, R., and Wharf, S.G. (1990) The effect of manipulation of salt pot hole size on table salt use. *Food Quality and Preference*, **2**, 13–20.

Giovanni, M.E. and Pangborn, R.M. (1983) Measurement of taste intensity and degree of liking of beverages by graphic scales and magnitude estimation. *Food Sci.*, **48**, 1175–82.

Hirsch, E. and Kramer, F.M. (1993) Situational influences on food intake, in *Nutrition for Work in Hot Environments*, National Academy Press, Washington DC, pp. 215–44.

Hirsch, E., Meiselman, H.L., Popper, R.D., Smits, G., Jezior, B., Lichton, I., Wenkam, N., Burt, J., Fox, M., McNutt, S., Thiele, M.N., and Direge, O. (1984) The effects of prolonged feeding Meal, Ready-to-Eat (MRE) operational rations. Technical Report Natick TR-85/035, US Army Natick Research, Development and Engineering Center, Natick, MA.

Jones, L.V., Peryam, D.R., and Thurstone, L.L. (1955) Development of a scale for measuring soldiers' food preferences. *Food Res.*, **20**, 512–20.

Lawless, H.T. (1984) Flavor description of white wine by 'expert' and nonexpert wine consumers. *J. Food Sci.*, **49**, 120–3.

Lawless, H.T. and Malone, G.J. (1986a) The discriminative efficiency of common scaling methods. *J. Sens. Studies*, **1**, 85–98.

Lawless, H.T. and Malone, G.J. (1986b) A comparison of rating scales: Sensitivity, replicates and relative measurement. *J. Sens. Studies*, **1**, 155–74.

Lester, L.S., Kramer, F.M., Edinberg, J., Mutter, S., and Engell, D.B. (1989) Evaluation of the Canteen cup stand and ration heater pad: effects on acceptability and consumption of the Meal, Ready-to-Eat in a cold weather environment. Technical Report Natick TR-90/008L, US Army Natick Research, Development and Engineering Center, Natick, MA.

McBride, R.L. (1982) Range bias in sensory evaluation. *J. Food Technol.*, **17**, 405–10.

McDaniel, M.R. and Sawyer, F.M. (1981) Preference testing of whiskey sour formulations: Magnitude estimation versus the 9-point hedonic. *J. Food Sci.*, **46**, 182–5.

Meilgaard, M., Civille, G., and Carr, B.T. (1987) *Sensory Evaluation Techniques*, Vols I and II, CRC Press, Boca Raton, Florida.

Meiselman, H.L. (1988) Consumer studies of food habits, in J.R. Piggott (ed.) *Sensory Analysis of Foods*, 2nd edn, Elsevier Applied Science, London.

Meiselman, H.L., Hirsch, E.S., and Popper, R.D. (1988) Sensory, hedonic, and situational factors in food acceptance and consumption, in D.M.H. Thomson (ed.) *Food Acceptability*, Elsevier Applied Science, London, pp. 77–87.

Morgan, T.E., Hodges, L.A., Schillig, D., Hoyt, R.W., Iwanyk, E.J., McAnich, G., Wells, T.C., and Askew, E.W. (1988) A comparison of the Meal, Ready-to-Eat, Ration, Cold Weather and Ration, Lightweight nutrient intakes during moderate altitude cold weather field training operations. Technical Report T5-89, US Army Research Institute of Environmental Medicine, Natick, MA.

Moskowitz, H.R., Kapsalis, J.G., Cardello, A.V., Fishken, D., Maller, O., and Segars, R.A. (1979) Determining relationships among objective, expert and consumer measures of texture. *Food Technol.*, **33**, 84–8.

Moskowitz, H.R. and Sidel, J.L. (1971) Magnitude and hedonic scales of food acceptability. *J. Food Sci.*, **36**, 677–80.

Pangborn, R.M., Guinard, J.X., and Meiselman, H.L. (1989) Evaluation of bitterness of caffeine in hot chocolate drink by category, graphic, and ratio scaling. *J. Sens. Studies*, **4**, 31–53.

Pearce, J.H., Korth, B., and Warren, C.B. (1986) Evaluation of three scaling methods for hedonics. *J. Sens. Studies*, **1**, 27–46.

Peryam, D.R. and Pilgrim, F.J. (1957) Hedonic scale method of measuring food preferences. *Food Technol.*, **11**(9), Suppl. 9–14.

Piggot, J.R. (ed) (1988) *Sensory Analysis of Foods*, Elsevier Applied Science, London.

Popper, R., Hirsch, E., Lesher, L., Engell, D., Jezior, B., Bell, B., and Matthew, W.T. (1987) Field Evaluation of Improved MRE, MRE VII, and MRE IV. Technical Report Natick TR-87/027, US Army Natick Research, Development and Engineering Center, Natick, MA.

Rogozenski, J.E. and Moskowitz, H.R. (1982) A system for the preference evaluation of cyclic menus. *J. Foodservice Systems*, **2**, 139–61.

Sampson, J. and Kobrick, J. (1980) Environmental symptoms questionnaire. Revisions and new field data. *Adviation, Space and Environmental Medicine*, **51**, 872–7.

Schutz, H.G. and Kamenetzky, J. (1958) Response set in measurement of food preference. *J. Applied Psychol.*, **42**, 175–7.

Shand, P.J., Hawrysh, Z.J., Hardin, R.T., and Jeremiah, L.E. (1985) Descriptive sensory assessment of beef steaks by category scaling, line scaling, and magnitude estimation. *J. Food Sci.*, **50**, 495–500.

Shepherd, R. and Griffiths, N.M. (1987) Preferences for eggs produced under different systems assessed by consumer and laboratory panels. *Lebensm. Wiss. u.-Technol.*, **20**, 128–32.

Shepherd, R., Griffiths, N.M., and Smith, K. (1988) The relationship between consumer preferences and trained panel responses. *J. Sens. Studies*, **3**, 19–35.

Stone, H., Sidel, J., Oliver, S., Woolsey, A., and Singleton, R.C. (1974) Sensory evaluation by quantitative descriptive analysis. *Food Technol.*, **29**(11), 24–34.

Turner, M. and Collison, R. (1988) Consumer acceptance of meals and meal components. *Food Qual. Pref.*, **1**(1), 21–4.

2 Appropriateness as a measure of the cognitive-contextual aspects of food acceptance

H.G. SCHUTZ

2.1 Introduction

Throughout the history of the scientific measurement of food acceptance, there has been a growing recognition that the use of preference alone, in the sense of pure affective judgement, is an insufficient measure as an explanatory variable for food choice. As evidenced by the data on the prediction of food consumption behavior illustrated by the work of Kamenetzky *et al.* (1957), Lau *et al.* (1979), Pilgrim and Kamen (1963), Sidel *et al.* (1972), the prediction of food consumption behavior from preference judgments at best accounts for only about 50% of the variance in consumption. The relevance of cognitive processes in food acceptance has been considered by Olson (1981), Thomson and McEwan (1988), Rappaport & Peters (1980), Worsley (1980). The consumer and market research community have also recognized the importance of so-called situational variables in understanding consumer behavior an indicated in the paper by Belk (1975), and in particular usage conditions as represented by Kakkar and Lutz (1981) and a pioneering work by Stefflre (1971). The use of Kelly's personal construct theory (1955) and repertory grid, illustrated by the work of McEwan and Thomson (1988) as one approach for evaluating the contextual aspects of food acceptance is discussed in detail by Gains in Chapter 3 of this volume. Although it is not the purpose of this chapter to argue for the use of one method of cognitive-context over another, it is relevant to indicate that the repertory grid procedure emphasizes the ideographic nature of context, that is, an individual's personal constructs and use of terms, whereas the appropriateness technique emphasizes the nomothetic approach, that is a common set of attributes evaluated by all respondents. The reader is advised to read the aforementioned chapter in this volume in order to develop a better sense of when one or the other procedure might be more appropriate.

As an historical note it was during the author's brief experience as a market researcher in 1970 working for Volney Stefflre that he first became exposed to the use of the 'item by use' appropriateness technique. After joining the University of California at Davis later in 1970 I saw the value of applying this generalized approach, not just to the goal of food product development, but as a tool for more completely understanding the cognitive structure of foods, as well as a methodology which might improve the predictability of food choice. Over the next 21 years the technique has been utilized to investigate food and food use classification (Schutz et al., 1975b) individual food classes: rice (Schutz et al., 1975a), wine (Schutz and Ortega, 1974), dairy products (Bruhn and Schutz, 1986) and vegetables (Resurreccion, 1986). Food choice and nutrition has been investigated (Baird and Schutz, 1976; Martens et al., 1988), and the prediction of food purchase and use by Schutz et al. (1977). Rucker and Schutz (1982) also utilized appropriateness data in developing a measure of consumer food typologies.

The use of this procedure in a hospital patient/employee study (Schutz et al., 1971) is a good example of where the procedure can be used for highly practical purposes. The dietitian of a hospital was interested in trying to develop menu items which would be more likely to be consumed by patients and thus contribute, from a nutritional standpoint, to their more rapid recovery. The determination of simple preferences taken out of context of the hospital situation would not provide such information, whereas obtaining appropriateness information from patients actually in the hospital situation allowed the collection of data which could be directly interpreted in guiding food preparation and menus. In addition, in this study it was possible to compare the patients' with the employees' reactions as another measure to help differentiate the importance of context in the acceptability of various foods.

Schutz (1988) has summarized most of the studies utilizing the appropriateness item by use technique and has made the case that the evidence is clear that measuring the cognitive aspects of food items represents an important non-affective dimension in understanding food choice for both theoretical purposes of classification as well as in practical applications in food product development and prediction of food use.

More recently, research has been conducted at the Institute of Food Research (IFR), Reading Laboratory by the author, in collaboration with IFR scientists, on the application of the technique to 50 representative foods in the UK, to food combinations, and to the investigation of the relationship of mood and food choice.

In spite of what is a rather obvious and important role for cognitive aspects of food in understanding food acceptance, remarkably little work has been done utilizing this concept. Simple global preference with all of its attendant problems continues to be the primary tool utilized by researchers in academia and industry. However with the growing interest in the consideration of food context we may well expect an increase in its utilization in the future.

2.2   Description of procedures for appropriateness, item by use technique

The remainder of the chapter will present in some detail the procedures for correctly utilizing the technique and, where deemed helpful, reference will be made to specific studies and their results.

The appropriateness, item by use technique in its simplest form represents a rating scale task in which respondents are asked to give a rating on appropriateness on an end anchored scale, from 1 (never appropriate) to 7 (always appropriate), for a set of food items and 'uses' which can cover a wide variety of situational conditions.

2.2.1   Selection of stimuli

2.2.1.1   Foods.   The food items chosen can either be broad, that is cover many classes of foods, or can represent individual food classes such as dairy products, wine, etc. In the author's experience the number of these food items can range from as few as 10 to as many as 58. Naturally the respondent burden is a consideration and will depend as well on the number of 'uses' for which the individual foods are rated. If the researcher's interest is primarily in understanding and classifying foods for a particular target group or culture, the broad goal (overall food classes) is an appropriate guideline for the selection of foods. In this case one must be cautious to select foods that represent all of the important dimensions in the culture, such as parts of meal, type of food (cereal, dairy, meat, etc.), foods which represent different sensory intensity levels, and be as much as possible representative of the foods in that particular culture (see Table 2.1 for an example of such a list for the USA).

If, on the other hand, interest is more in a particular food class, such as dairy products or vegetables, in some ways it is a simpler choice because of the limited number of items possible in a particular class. However, if the objective is to understand that class and products in relationship to a marketing and/or food product development objective, it is important to have, in addition to members of that food class, representatives of other foods which are substituted or considered surrogates by the consumer. For example, in the study by Schutz and Ortega (1974) on wine, other beverages including hard liquor, beer, and carbonated beverages are included. (See Table 2.2 for the list of beverages in this study.) The choice of these items is not guided by any hard and fast rules, but rather by such factors as frequency of use, share of market and representativeness for particular classes of surrogate products.

A natural question which arises in the selection of food items is the degree of specificity of the naming of items that are selected. Thus, does

Table 2.1 List of foods

Pie	Yogurt
Potato salad	Roast beef
Milk	Spaghetti
Tomatoes	Fried eggs
Chicken	Tossed salad
Coffee	Orange juice
Jello	Ice cream
American cheese	Baked beans
Shrimp	Rice
Tea	Carrots
Chili	Bagels
Vegetable soup	Broccoli
Liver	Cottage cheese
Fish	Peanut butter
Soft drinks	French fries
Meat loaf	Ham
TV dinners	Peas
Watermelon	Strawberries
Steak	Potato chips
Wine	Tuna
Dry cereal	Pizza
Cake	Frankfurters
Dip	Pickles
Chop suey	Onions
Apples	Parsley
Chitterlings	Candy bars
Bread	Bacon
Hamburger	Tacos

(Adapted from Schutz *et al.*, 1975b.)

one list eggs, fried eggs, or use the brand name for an item? Obviously these are likely to be important factors in the way in which the respondent views stimuli. The examination of such considerations in previous research indicates that differences in preparation procedure or the use of various modifiers such as brand does make a difference in the nature of appropriateness ratings. Thus the researcher must decide which level of specificity is most meaningful for their research objectives, and in some cases it may be necessary to use several levels in order to understand the perceptual differences that may exist.

Another question which may arise in the reader's mind is: Can such a method be utilized for actual food stimuli rather than just for the names of food? Research on this topic is being undertaken by the author at the present time. There seems no reason why at least a simplified form of an item by use matrix could not be utilized as a part of laboratory food sensory evaluation, central location or in-home testing. Just as with attitudinal data collected on names of food it would add an important dimension to the understanding of foods from a theoretical perspective as well as from the standpoint of food product development guidance.

Table 2.2 List of questionnaire items

Whiskey	A dry wine
A red wine	Soft drink (soda pop)
Burgundy	Scotch
A green bottle of wine	'Sémillon'
Rhine wine	A sweet wine
Chablis	A half-bottle of wine (tenth)
Sauterne	Slightly sparkling wine
'Zinfandel'	'Cabernet'
Fruit juice	Punch
A white wine	Wine in a Burgundy bottle
Cold duck	Coffee
Sherry	'Chenin blanc'
Extra-dry champagne	A fifth-bottle of wine
Gin	A dessert wine
A half-gallon bottle of wine	Cocktails
Apple wine	'Gamay' rosé
Imported wine	A brown bottle of wine
'Riesling'	A table wine
'Pinot noir'	'Chardonnay'
A clear bottle of wine	Liqueur
Vodka	Tea
'Gamay'	A rosé wine
'Green hungarian'	'Pinot blanc'
Port	Bourbon
A magnum bottle of wine	A gallon-bottle of wine
'Sauvignon blanc'	An aperitif wine
'Grenache' rosé	Brut champagne
Fruit-flavored wine	Beer

(Adapted from Schutz and Ortega, 1974.)

As was mentioned earlier the approach has been utilized most recently in looking at food combinations, thus a food item in addition to being a single food could be a food combination such as bacon and eggs or even a list of items that could compose an entire meal.

2.2.1.2   Uses.   As with food items the selection of uses is influenced by the objectives of the particular research. However an overriding guideline is that the uses cover all of the representative and frequent uses for the type of product that is involved. Can there be one set of uses that can be applied across all foods that represents all the major use situations for individuals? In looking back at previous work there do seem to be some uses that appear in most of the studies, however they are not identical.

Table 2.3 presents an example of basic usage situation variables that could be considered in the development of a list (also see Table 2.7 for another list). As is obvious from examination of these tables, it is not difficult to develop very quickly a large number of usage situations; a more significant problem is limiting the number so as not to increase the burden on the respondent. It is not suggested that it is necessary to

Table 2.3 Uses – attributes for wine and related products

Time of day	*Psychological*
Goes well with meals	When I'm sad
For cocktail hour	When I want something different
To drink before bed	To drink when celebrating
Something that's good anytime	Something I really like
As an evening snack	When I want to feel creative
Good with lunch	Promotes sociability
	A relaxing drink
Occasion served	To impress someone
To serve when entertaining	Gives feeling of well-being
For special holidays	
At parties	*Person served*
	For men
Where served	For the elderly
To use at home	For guests
When I'm at a restaurant	To drink at any age
	For young adults
Physiological	For women
As an aid to indigestion	
To help me sleep	*Physical*
Something good for me	A refreshing drink
When I'm watching my weight	As part of daily meals
A stimulating drink	A drink that is soft
Picks up my appetite	A fragrant drink
When I'm not feeling well	Easily stored
	A light drink
How used	
To drink straight	
Mixes well in cocktails	
As an appetizer (aperitif)	
Goes well with fish or poultry	
An after dinner drink	
Good over ice	
Goes well with meat	
To cook with	
Hard to serve correctly	

(Adapted from Schutz and Ortega, 1974.)

have uses from all of these categories. Rather they must be selected, as indicated earlier, on the basis of the objectives of the particular research study. However, the more global one's interest is in understanding the nature of the cognitive aspects of foods, the more one is required to include representatives of the various dimensions that are listed, just as in the choice of food items.

Borrowing from the realm of consumer behavior we recently have included psychological variables which describe degree of involvement such as 'something I spend a lot of time thinking about'. This type of variable might be expected to have special significance in the prediction of food consumption. Preliminary support for this hypothesis has been found in recent studies conducted at IFR (Shepherd *et al.*, in prep.).

Expression of uses varies from the simple statement, for example, 'for women', to ones where it seems more accurate to utilize a phrase such as 'something I really like', a 'food I really like', or 'when I want something I really like'. Again there is no research that demonstrates the possible difference in results that one might obtain with different methods of expression. However two factors have guided the selection in previous studies. First, to keep the statement as short and as simple as possible and second, to have it reflect more accurately the particular use situation.

Again as with the food items, the number of uses that one selects depends on the purpose of the study, but the combinations of foods and uses grow geometrically, so that a 10 by 10 is 100 judgments, whereas a 50 by 50 represents 2500 judgments on the part of the respondent. The author has utilized both extremes and they both work, however it must be obvious that with the longer task individuals do not complete the questionnaire in one sitting. A longer questionnaire, such as 50 by 50, can take up to three hours for completion.

For most of the previous studies in which the author has been involved, one food use was included which can be considered as a surrogate measure of preference, typically stated as 'a food I really like'. The face validity of this attribute as a measure of hedonic value seems reasonable. However one might argue that in the context of the many other attributes measurements that are made in this type of study, results obtained with a separate single measure of hedonic value might provide slightly different results, although it is hard to believe that the basic order of the results would be much changed if the affective measure were measured independently.

If the researcher does not have a selection of food items or uses that can be derived from the literature or their specific knowledge of an area, there is an experimental procedure for generating a number of alternatives. Individual respondents sampled from the population of interest are asked to name items from the class of foods involved, then asked what they can be used for or by whom (uses/situations), and then asked what other items could be used for that purpose. If 10 to 20 respondents are interrogated in this manner a large list of food items and uses, from a consumer point of view, can be developed. This list plus other alternatives selected by the researcher can make up a larger list from which the final selection of food items and uses can be made.

2.2.2 Format of the questionnaire

The basic format which has been utilized in appropriateness research is illustrated in Figure 2.1. This is a simple matrix format with items listed on one axis (usually vertical) and uses along the other (usually horizontal) with the instructions to the respondent to put the appropriateness ratings within each cell. It has been the practice to have respondents rate all of

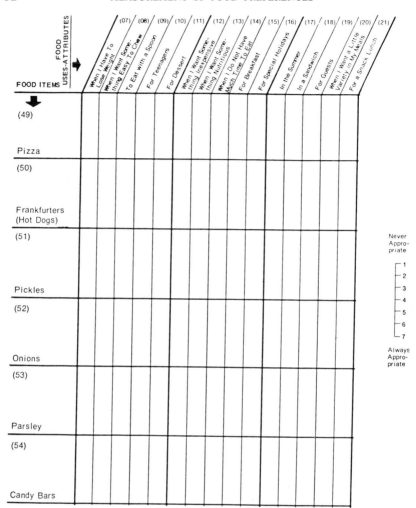

Figure 2.1 Example of item by use format.

the uses for a food and then go on to the next food rather than have a use rated over all foods and then go on to the next use. Frankly the author has never studied the influence of rating in this latter manner, however, it always seemed more reasonable to have the respondent think about the particular food and all of its uses than to have the respondent be comparing foods over all uses. A simple experiment to test the possible differences in the two approaches might be interesting. In the case where the number of items and/or uses is greater than the number than would fit into a matrix of one page, two approaches have been utilized. One is to create a number of matrix pages in which all the uses appear across the top and are constant and the foods vary from page to page along the vertical axis. Obviously there are as many pages as necessary to include all of the food items to be studied. Another format, which will work, has multiple pages in which on the side of the pages the various uses are repeated and four or five food items are listed along the top. Again the number of pages is determined by the number of food items that are involved. As the reader may note, in the former matrix situation the respondent is rating across the page and in the latter, down the page. To the best of my knowledge this variation in rating direction does not produce any problem for the respondents or difference in results.

The number of points on the scale of appropriateness has been investigated (Schutz and Rucker, 1975) where 2, 3, 6, and 7 points were examined, with the results that the basic cognitive structure was identical for all scales. This would seem to indicate a high degree of robustness in the type of measurements that are made. One study (Baird and Schutz, 1976) utilized just an appropriate/inappropriate judgment (0, 1) thus a 2 point scale, and included two languages other than English, Spanish and Chinese, with meaningful results.

One issue of concern with regard to instructions to the respondents in their rating task is that concerning unfamiliar foods. For at least some respondents the food items that are listed may represent ones which they either have never heard of, or cannot remember consuming. It has been the practice to instruct respondents to make their ratings anyway, and rate what they think would be the best representation for each food use combination. We then ask the respondent to circle a food in this manner. Analysis of results with and without the inclusion of the unfamiliar foods or between familiar and unfamiliar means shows no significant difference in scores. One hypothesis to explain this result is that subjects would tend to rate these foods closer to the middle point of the scale and secondly that even though the respondent may not have consumed the food, information relative to the cultural stereotypes for the use of this food are known to the individual and therefore reflected in their ratings. Respondents in general are instructed not to spend much time thinking about responses but to go quickly through the rating process giving the first response that comes to mind.

2.2.3 Selection of respondents

Depending on the population to which one wishes to generalize the results and the purpose of the study, the selection of subjects can range from a rather narrow target population, such as young female users of a product or brand, to a random selection of the general population.

The sampling procedure may be one which is a convenience/judgment type sample, a random sample, or it is not uncommon to use some type of quota sampling to ensure enough respondents in the important strata for the particular objectives of the study. How many people do you need to conduct such a study in order to have reliable and valid results? The results to date are remarkably consistent in that as few as 25 respondents, in whatever class in which one is interested, can produce reliable means for an item by use pair and consistent cognitive structures. Thus one might utilize as few as 25 respondents if no further breakdown of the subject population were contemplated, although it has been typical to utilize around 50. Obviously if one wanted to develop very small standard errors of the mean for purposes of testing the statistical significance among various pairs of items, a larger N would be necessary. Also, if one is going to break out the population into subgroups whether by demographic or other attributes, a larger N will be necessary.

Why is it that such a small number of respondents can yield meaningful results when compared to hedonic type or some other market research judgments where larger Ns are necessary in order to obtain such reliability? I believe it is the basic nature of the difference between cognitive and affective judgments. All of the various factors which influence hedonic judgments can and usually do operate when the respondent is making a judgment. Thus previous consumption, preparation, and all the other contextual components possible go in to create a high level of heterogeneity across respondents. On the other hand, for the appropriateness judgment, one obtains a more culturally common type of judgment. That is, individuals operate in their judgments in a more consistent way with what is typical and accepted in their culture. Thus one achieves much more homogeneity in judgements across people in a particular population. One might think that there would be a wide variation in affective judgments if one were to ask a group of respondents how much they liked turtle soup. On the other hand, there would probably be quite a consistent response if one were to ask the same respondents how appropriate they thought turtle soup was to eat 'when riding in a car', etc.

Respondents that have been utilized have included members of both genders, the use of three cultures (and languages – Spanish, Chinese, Norwegian) other than English, adults from 18–65, but no individuals under 18. There is no reason to believe that the procedure could not be used with children especially if one were to use a simple two-point scale perhaps called 'OK' and 'not OK'.

2.2.4   Collection of non-appropriateness data

In addition to the item by use matrix the researcher may wish to collect, as has been done in many studies, other types of data which may be used in several ways: as a way of classifying people for subgroup analysis; to check the representativeness of the population that has been studied; to relate to appropriateness ratings; to predict food consumption. Thus data on demographics can be collected of a standard nature, including age, income, gender, household size, occupation, specific dieting behavior, etc. If one is interested in developing a concurrent measure of food consumption to use for prediction purposes, a simple frequency of consumption measure can be obtained for all or selected items that are in the study. Data on particular brand usage or other market specific information can be useful if the objective of the research includes product development. Naturally any other information specific to the purpose of the study may be collected within the general guidelines of producing a questionnaire which is not unnecessarily burdensome for the respondent.

2.2.5   Data collection

The data collection procedures are dependent in part upon the length of time required to complete the questionnaire, in part the speed with which the results are needed, and resources available (people, money, etc.). For a shorter questionnaire, a mail or perhaps even a telephone data (probably no more than a 10 by 10 matrix) collection technique can be utilized. For the longer questionnaires we have always utilized a technique which combines in-home with mail back or pickup procedures. The questionnaire is delivered, instructions given, and the questionnaire left to be filled out over a week-long period of time. Then it is either mailed back or picked up by the experimenter at the end of a week. Return rates can be expected to range from a low of 50% (for four wave mail procedure) to a high of around 95% for an in-home delivered questionnaire.

For mail questionnaires the author has not given rewards to respondents to act as an incentive to fill out and return the questionnaire. For the three hour in-home procedure, rewards should be given and in the past have ranged from money ($5–10) to store coupons of equivalent value.

Regardless of which data collection is used, instructions for rating must be carefully constructed and well communicated. In some instances an example has been used to help clarify the rating task.

2.2.6   Analysis of appropriateness data

At the simplest level computing the means over the total sample for each item by use pair is the first step. Frequency distributions and standard deviations are also useful to ascertain whether or not there are any strongly skewed distributions and for use in testing for significant differences between means, if so desired. If a food item shows little variation for certain uses it may be appropriate to drop that item or at least be aware of its potential influence on subsequent analyses.

If missing data are found, if few in number ($< c.$ five) they can be replaced by the mean for that item by use pair over the remaining respondents. If greater than five, the use of a multiple regression approach for missing data should be considered. A question arises as to whether or not a rating scale of this type (category) is one which can utilize a central tendency measure such as the mean (alternately one can also compute a 'percent appropriateness value', percent of the total of 5, 6, and 7 ratings). Theoretical and practical experience with ordered metric scales of this type in a wide variety of applications including the appropriateness methodology indicates that it can be analysed as if it were a parametric measure such as that obtained with a true interval scale.

As the reader might imagine, with a large matrix such as 50 by 50 a 'generous' amount of data is produced from just a simple mean analysis. It is useful to have a computer program construct mean tables in which for each item the use attributes are printed out in order from the highest to lowest mean, as well as for the uses to have the foods means printed out from the lowest to highest mean. This basic set of tables, which for a large matrix will never see the right of publication because of its voluminous nature, can be of a very practical value in working out food product development and positioning questions. A way of presenting a representative, if not comprehensive, set of means for foods and uses will be covered later in the chapter.

When examining these tables one should remember that these are perceptions, not facts, and in some instances may differ widely from expert judgments or objective data. This type of discrepancy can be useful in identifying areas where consumer education (advertising) may be desirable.

For many studies using the appropriateness methodology it is desirable to have a way of grouping or organizing the foods and/or uses to better understand the relationships among the various variables. Computing correlations between each pair of foods with the uses as cases is a standard approach for looking at relationships between pairs. This can be done for the uses as well where the foods serve as cases. If the researcher is interested in a measure of relationship which indicates absolute similarity of measure and not simply similarity of pattern as with a Pearson product–moment correlation, a useful measure is scale distance (D) (Rummel, 1970). The author has used both the scaled distance and the correlation coefficient and also compared the two and found them remarkably similar in terms of the conclusions with which one would draw from the data. Tables 2.4 and 2.5 illustrate scaled distances of selected food items and uses respectively. The benefit of the scale distance

Table 2.4 Scaled distances[a] between factor-defining[b] foods

	Factor 1 foods			Factor 2 foods			Factor 3 foods			Factor 4 foods			Factor 5 foods		
	Wine	Pie	Cake	Liver	Chili	Chitterlings	Chicken	Roast beef	Steak	Jello	Cottage cheese	Orange juice	Peanut butter	Candy bars	Potato chips
Wine	1.00	0.70	0.66	0.62	0.64	0.62	0.58	0.62	0.60	0.60	0.60	0.63	0.58	0.63	0.66
Pie		1.00	0.88	0.62	0.64	0.55	0.62	0.68	0.64	0.67	0.64	0.62	0.60	0.67	0.66
Cake			1.00	0.58	0.60	0.50	0.65	0.67	0.61	0.68	0.63	0.62	0.61	0.69	0.71
Liver				1.00	0.77	0.71	0.63	0.67	0.71	0.56	0.66	0.61	0.64	0.60	0.55
Chili					1.00	0.70	0.59	0.63	0.63	0.59	0.64	0.60	0.66	0.64	0.59
Chitterlings						1.00	0.45	0.48	0.49	0.46	0.54	0.53	0.62	0.62	0.52
Chicken							1.00	0.80	0.74	0.63	0.65	0.60	0.60	0.59	0.64
Roast beef								1.00	0.82	0.61	0.64	0.61	0.60	0.57	0.61
Steak									1.00	0.56	0.60	0.60	0.55	0.53	0.56
Jello										1.00	0.79	0.70	0.61	0.61	0.60
Cottage cheese											1.00	0.72	0.66	0.61	0.59
Orange juice												1.00	0.68	0.67	0.65
Peanut butter													1.00	0.73	0.67
Candy bars														1.00	0.78
Potato chips															1.00

[a] These values range from 0 to 1 with 1 being identity.
[b] The three foods with the highest loadings on a factor are considered factor-defining foods.
(Adapted from Schutz et al., 1975b.)

Table 2.5 Scaled distances[a] between factor-defining[b] uses

	Factor 1 uses			Factor 2 uses			Factor 3 uses			Factor 4 uses		
	Teens	Children	Easy	Unhappy	Riding	Not well	Main dish	Fork	Dinner	Parties	Friends	Guests
For teenagers	1.00											
For children	0.89	1.00										
Easy to prepare	0.78	0.78	1.00									
When unhappy	0.40	0.45	0.52	1.00								
Riding in a car	0.34	0.39	0.45	0.84	1.00							
Not feeling well	0.39	0.44	0.52	0.84	0.75	1.00						
As a main dish	0.44	0.49	0.49	0.61	0.55	0.57	1.00					
Eat with a fork	0.54	0.56	0.55	0.48	0.40	0.44	0.62	1.00				
For dinner	0.64	0.65	0.67	0.53	0.43	0.50	0.65	0.71	1.00			
At parties	0.58	0.58	0.66	0.64	0.56	0.58	0.53	0.52	0.64	1.00		
With friends	0.68	0.68	0.75	0.61	0.52	0.56	0.55	0.56	0.71	0.85	1.00	
For guests	0.69	0.68	0.74	0.53	0.44	0.49	0.49	0.57	0.72	0.80	0.86	1.00

[a] These values range from 0 to 1 with 1 being identity.
[b] The three uses with the highest loadings on a factor are considered factor-defining uses.
(Adapted from Schutz et al., 1975b.)

procedure occurs when one is interested in actually substituting one item for another rather than for just looking at relationships. For example, if such a study were being done in order to determine the likelihood of substituting one food from one culture to another it would be important to be able to look at absolute similarity rather than just similar patterns in order to help ensure the substitutability of one food for another.

2.2.7   Use of principal component analysis

One of the earliest uses of the appropriateness technique was an attempt to better understand the way in which consumers organize and classify foods and uses. It was our belief that consumers have a simpler perceptual structure to foods than individual items and that the classification or structure suggested by anthropologists or nutritionists did not necessarily represent the one that existed in the minds of consumers. To develop these simpler structures the approach utilized principal component analysis (PCA). As a method of factor analysis it is the least influenced by subjective judgment and can utilize a wide variety of relationship data such as correlations or scaled distance. The author does not mean to imply that other grouping techniques such as other methods of factor analysis or cluster analysis could not be utilized if one is interested in looking at groupings or common dimensions obtained from appropriateness type data. It is that this is what has been used in previous research and it can be unequivocally stated that this type of analysis results in dimensions which account for meaningful factors and a high variation in relationships among variables (from 70 to 90%). It is not the purpose of this paper to explicate the details of conducting this type of analysis since that is available from many other sources. However it has been standard in the author's research to utilize an eigen value of 1 as a stopping point for factoring, and Varimax rotation (an orthogonal structure) as a method for obtaining the most simple and meaningful psychological structure. The correlation input matrices used for foods, where the uses are the cases, ends up with food factors, while the principal component analysis on uses with the inverse correlation matrix of uses with foods as cases, yields usage factors. The factors have a theoretical meaning but also have practical value if one is interested in ways of displaying results or for helping to select foods or uses for other studies.

The PCA data for a general foods study (Schutz et al., 1975b) is displayed in Table 2.6 for food items and Table 2.7 for uses in which the factor weights are given from high to low and displayed only for those for the factor weight exceeding 0.50. The variance accounted for by each factor as well as for the total of factors and the communalities are typically presented. The naming of factors, as is the case in all PCAs, is

Table 2.6 Food factors with rotated factor loadings[a], communalities, and proportion of variance accounted for by each factor

	Total sample					
	Factor 1 High-calorie treat	*Factor 2* Specialty meal items	*Factor 3* Common meal items	*Factor 4* Refreshing healthy food	*Factor 5* Inexpensive filling foods	
Food						*Communalities*
Wine	0.63	–	–	–	–	0.76
Pie	0.63	–	–	–	–	0.79
Cake	0.62	–	–	–	–	0.77
Dip	0.61	–	–	–	–	0.78
Soft drinks	0.57	–	–	–	–	0.80
Potato chips	0.52	–	–	–	0.55	0.81
Ice cream	0.52	–	–	0.58	–	0.78
Strawberries	0.50	–	–	0.57	–	0.82
Candy bars	–	–	–	–	0.56	0.79
Pickles	–	–	–	–	0.50	0.82
Bagels	–	0.53	–	–	–	0.83
Onions	–	0.57	–	–	–	0.75
Watermelon	–	–	–	0.51	–	0.76
Coffee	–	–	–	–	–	0.63
Tacos	–	0.69	–	–	–	0.84
Parsley	–	0.63	–	–	–	0.74
Pizza	–	0.50	–	–	–	0.74
Chitterlings	–	0.75	–	–	–	0.76
Tea	–	–	–	0.57	–	0.74
Shrimp	–	–	0.59	–	–	0.77
Liver	–	0.75	–	–	–	0.84
Chili	–	0.74	–	–	–	0.80
TV dinner	–	0.73	–	–	–	0.78
Chop suey	–	0.73	–	–	–	0.87
Baked beans	–	0.71	–	–	–	0.84
Peas	–	0.69	–	–	–	0.84
Rice	–	0.67	–	–	–	0.86
Broccoli	–	0.66	–	–	–	0.83
Spaghetti	–	0.64	0.51	–	–	0.80
French fries	–	0.62	–	–	–	0.79
Meat loaf	–	0.60	0.58	–	–	0.84
Fried eggs	–	0.60	–	–	–	0.78
Fish	–	0.58	0.62	–	–	0.84
Vegetable soup	–	0.58	–	0.51	–	0.75
Bacon	–	0.52	–	–	–	0.78
Dry cereal	–	0.50	–	0.63	–	0.80
Potato salad	–	–	–	–	–	0.76
Yogurt	–	–	–	0.64	–	0.80
Carrots	–	–	0.50	–	–	0.80
Chicken	–	–	0.76	–	–	0.83
Roast beef	–	–	0.74	–	–	0.82
Steak	–	–	0.69	–	–	0.76
Ham	–	–	0.69	–	–	0.82
Hamburger	–	–	0.69	–	–	0.82
Tomatoes	–	–	0.64	–	–	0.80
Tossed salad	–	–	0.60	–	–	0.79
Tuna	–	–	0.65	–	–	0.81
Franks	–	–	0.57	–	–	0.81
American cheese	–	–	0.56	–	0.50	0.85
Bread	–	–	–	–	0.55	0.78
Jello	–	–	–	0.70	–	0.82

Table 2.6 *Continued*

| | Total sample | | | | | |
| | *Factor 1* High-calorie treat | *Factor 2* Specialty meal items | *Factor 3* Common meal items | *Factor 4* Refreshing healthy food | *Factor 5* Inexpensive filling foods | |
Food						*Communalities*
Cottage cheese	–	–	–	0.67	–	0.82
Orange juice	–	–	–	0.64	–	0.77
Milk	–	–	–	0.62	–	0.80
Peanut butter	–	–	–	–	0.57	0.80
Apples	–	–	–	0.52	–	0.74
Proportion of variance accounted for by each factor and total variance	0.13	0.23	0.19	0.15	0.10	0.80

[a] Only factor loadings of 0.50 and above are included.
(Adapted from Schutz *et al.*, 1975b.)

done by determining the basic nature of the highly weighted variables on each factor and then finding a simple term (if possible) to describe that dimension.

With the availability of graphic capability from statistical computer programs it is possible to display in graphic form the relationship between any two factors, and if desired the vectors for the uses of each of these factors.

If analyses are conducted for subgroups of the population such as age, gender, or geographical area, PCAs can be conducted for each and various methods of factor comparison can be conducted. These can be as basic as comparing the number of factors and their composition, a comparison of the variance accounted for by comparable factors as well as total variance, and if desired a more sophisticated approach using Procrustes analysis (Gower, 1975) where one can obtain a more quantitative estimate of the degree of similarity or differences among factor structures for different groups.

As mentioned earlier it is possible to use the grouping analysis as an aid in helping to display meaningful and representative pairs of means from the data. For example, it is convenient to select 3 to 5 of the highest weighted foods on each factor and 3 to 5 of the highest weighted uses on each factor and utilize them to construct a simplified display of appropriateness means which are helpful in understanding the basic results of the study. An example of this is shown in Table 2.8 for foods and Table 2.9 for uses based on the results from the study on general food classification (Schutz *et al.*, 1975b).

From examination of these tables it is possible to get a quick overview of the basic findings of the study. In an even simpler format only the top

Table 2.7 Use factors with rotated factor loadings[a], communalities, and proportion of variance accounted for by each factor

Food	Total sample				
	Factor 1 Utilitarian	Factor 2 Casual	Factor 3 Satiating	Factor 4 Social	Communalities
For teenagers	0.85	–	–	–	0.79
For children	0.83	–	–	–	0.79
Easy to prepare	0.83	–	–	–	0.83
In summer	0.83	–	–	–	0.80
For men	0.79	–	–	–	0.80
Easy to digest	0.77	–	–	–	0.83
Nutritious	0.77	–	–	–	0.79
For lunch	0.76	–	–	–	0.81
Inexpensive	0.75	–	–	–	0.79
On cold days	0.74	–	–	–	0.79
By itself	0.73	–	–	–	0.78
Not much time	0.72	0.55	–	–	0.84
Easy to chew	0.71	–	–	–	0.81
Really like	0.71	–	–	–	0.86
For guests	0.65	–	–	0.57	0.86
Not to run out of	0.65	0.53	–	–	0.73
With friends	0.62	–	–	0.58	0.89
For dinner	0.61	–	0.54	–	0.80
Visiting	0.58	–	–	0.56	0.88
Served cold	0.57	–	–	–	0.66
Really hungry	0.57	–	0.52	–	0.83
Eating out	0.55	–	–	–	0.81
Variety	0.55	–	–	–	0.81
Something light	0.53	0.73	–	–	0.83
Not very hungry	0.50	0.76	–	–	0.85
Lose weight	0.50	0.55	–	–	0.63
Unhappy	–	0.83	–	–	0.85
Riding in car	–	0.83	–	–	0.78
Not feeling well	–	0.82	–	–	0.79
For dessert	–	0.76	–	–	0.67
With cocktails	–	0.75	–	–	0.75
Eat with spoon	–	0.73	–	–	0.64
For breakfast	–	0.74	–	–	0.64
Watching TV	–	0.73	–	–	0.81
Something you broil	–	0.72	–	–	0.74
Between-meal snack	–	0.70	–	–	0.84
With coffee	–	0.70	–	–	0.74
Snack lunch	–	0.68	–	–	0.68
Feel creative	–	0.68	–	–	0.80
A spicy food	–	0.67	–	–	0.72
In a salad	–	0.65	–	–	0.61
Eat with fingers	–	0.64	–	–	0.61
In a sandwich	–	0.57	–	–	0.59
On a picnic	–	0.53	–	–	0.69
As a main dish	–	–	0.66	–	0.76
Eat with fork	–	–	0.60	–	0.67
At parties	–	–	–	0.65	0.86
For special holidays	–	–	–	0.57	0.81
Proportion of variance accounted for by each factor and total variance	0.31	0.30	0.08	0.08	0.77

[a] Only factor loadings of 0.50 and above are included.
(Adapted from Schutz et al., 1975b.)

Table 2.8 Selected[a] mean appropriateness ratings for foods[b]

	Factor 1 foods		
Uses	Wine	Pie	Cake
Something you broil	1.04	1.04	1.06
Riding in a car	1.04	1.22	1.84
For breakfast	1.04	1.50	1.83
In a sandwich	1.08	1.03	1.06
As a main dish	1.18	1.22	1.10
A spicy food	1.30	1.66	1.46
In a salad	1.64	1.02	1.04
Have to lose weight	1.77	1.06	1.24
Not feeling well	1.98	1.54	1.54
With cocktails	–	1.16	1.40
For guests	6.46	6.52	6.58
For special holidays	6.44	6.52	6.32
At parties	6.36	–	6.22
With friends	6.08	–	6.18
For dessert	–	6.64	6.68
Eat with a fork	–	6.54	6.35
With coffee	–	6.36	6.32

	Factor 2 foods		
Uses	Liver	Chili	Chitterlings
Riding in a car	1.06	1.07	1.26
For dessert	1.07	1.06	1.13
Eat with fingers	1.15	1.12	1.71
For breakfast	1.23	1.08	1.28
In a salad	1.23	1.16	1.24
On a picnic	1.26	1.57	1.34
Watching TV	1.34	1.74	1.43
For a snack	1.37	1.86	1.46
When unhappy	1.42	1.72	1.27
Served cold	1.49	1.16	1.48
In a snack lunch	1.52	1.64	1.32
Not feeling well	1.58	1.28	1.17
With cocktails	1.67	1.36	1.32
Something light	1.79	1.69	1.30
For special holidays	1.80	1.74	1.57
Eat with a spoon	1.12	–	1.47
At parties	1.78	–	1.48
Not hungry	1.80	–	1.47
A spicy food	1.83	–	1.80
Try not to run out	1.96	–	1.15
Have to lose weight	–	1.49	1.29
In a sandwich	–	1.40	1.16
Something you broil	–	1.16	1.29
For dessert	1.06	1.03	1.06
Eat with a spoon	1.19	1.10	1.06
For breakfast	1.12	1.26	–
Riding in a car	–	1.56	1.10
Eat with fingers	–	1.90	1.26
In a salad	–	1.68	1.17
For dinner	6.60	6.75	6.88

Table 2.8 *Continued*

| | Factor 3 foods | | |
Uses	Chicken	Roast beef	Steak
As a main dish	6.57	6.74	6.70
For teenagers	6.43	6.42	6.25
For men	6.38	6.76	6.79
Something nutritious	6.31	6.52	6.56
For guests	6.27	6.66	6.42
For children	6.37	6.30	–
In summer	6.33	–	6.21
In a sandwich	6.21	6.40	–
Eat with a fork	–	6.51	6.78
On cold days	–	6.39	6.40
Really hungry	–	6.36	6.55
Something I really like	–	6.31	6.62
When eating out	–	6.22	6.56

| | Factor 4 foods | | |
Uses	Jello	Cottage cheese	Orange juice
Something you broil	1.08	1.02	1.08
In a sandwich	1.08	1.52	1.05
Eating with fingers	1.17	1.10	1.04
Riding in a car	1.18	1.10	1.46
A spicy food	1.22	1.18	1.12
A main dish	1.57	–	1.09
With cocktails	1.15	1.51	–
Served cold	6.72	6.64	6.87
In summer	6.45	6.26	6.60
For children	6.56	–	6.63
For teenagers	6.49	–	6.48
Easy to digest	6.48	6.26	–
Easy to chew	6.22	6.22	–
Something nutritious	–	6.20	6.56

| | Factor 5 foods | | |
Uses	Peanut butter	Candy bars	Potato chips
Something you broil	1.06	1.05	1.06
A spicy food	1.14	1.20	1.78
In a salad	1.15	1.08	1.51
Eat with a fork	1.22	1.15	1.16
A main dish	1.30	1.09	1.20
Have to lose weight	1.40	1.17	1.14
Not feeling well	1.45	1.43	1.46
Want to feel creative	1.54	1.46	1.89
When eating out	1.30	1.25	–
With cocktails	1.34	1.19	–

Table 2.8 *Continued*

	Factor 5 foods		
Uses	Peanut butter	Candy bars	Potato chips
For dinner	1.42	1.26	–
For dessert	1.59	–	1.68
Eat with a spoon	–	1.08	1.06
For breakfast	–	1.08	1.14
In a sandwich	–	1.08	1.46
With coffee	–	1.85	1.67
For teenagers	6.14	–	6.18
Eat with fingers	–	6.35	6.50

[a] Only consistent extreme means are reported, i.e., at least two out of the three factor-defining uses for each factor must be rated never appropriate (1.00 to 1.99) or always appropriate (6.00 to 6.99) for a given use.
[b] The three uses with the highest loadings on a factor are considered factor-defining foods.
(Adapted from Schutz *et al.*, 1975b.)

Table 2.9 Selected[a] mean appropriateness ratings for uses[b]

	Factor 1 uses		
Foods	For teenagers	For children	Easy to prepare
Wine	1.44	1.16	–
Chitterlings	–	1.89	1.48
Ice cream	6.64	6.58	6.02
Frankfurters	6.46	6.33	6.18
Milk	6.63	6.79	–
Hamburgers	6.62	6.49	–
Jello	6.49	6.56	–
Orange juice	6.48	6.63	–
Spaghetti	6.46	6.50	–
Apples	6.45	6.63	–
Chicken	6.43	6.37	–
Roast beef	6.42	6.30	–
Watermelon	6.36	6.14	–
American cheese	6.32	6.17	–
Bread	6.30	6.44	–
Dry cereal	6.24	6.30	–
Peanut butter	6.14	6.50	–
Vegetable soup	6.10	6.18	–

	Factor 2 uses		
Foods	When unhappy	Riding in a car	Not feeling well
Chitterlings	1.27	1.26	1.17
Parsley	1.28	1.17	1.26
Onions	1.39	1.31	1.08
Liver	1.42	1.06	1.58
Peas	1.42	1.08	1.50
Baked beans	1.44	1.09	1.30

Table 2.9 *Continued*

Foods	Factor 2 uses		
	When unhappy	Riding in a car	Not feeling well
Broccoli	1.48	1.07	1.62
Chili	1.72	1.07	1.28
Carrots	1.72	1.77	1.82
Potato salad	1.76	1.14	1.31
Meat loaf	1.80	1.20	1.84
Fish	1.82	1.20	1.80
Tomatoes	1.83	1.53	1.68
Tuna	1.84	1.52	1.60
Chop suey	1.89	1.09	1.30
Pickles	1.89	1.87	1.22
Tacos	1.90	1.65	1.12
Peanut butter	1.94	1.88	1.45
Dry cereal	1.58	1.40	–
Rice	1.66	1.08	–
Fried eggs	1.68	1.11	–
TV dinners	1.72	1.06	–
Yogurt	1.90	1.24	–
Bagels	1.80	–	1.68
Wine	–	1.04	1.98
Spaghetti	–	1.10	1.55
Dip	–	1.12	1.14
Pie	–	1.22	1.53
Bacon	–	1.25	1.63
Watermelon	–	1.28	1.88
Shrimp	–	1.34	1.66
Ham	–	1.69	1.60
Cake	–	1.84	1.54
Pizza	–	1.93	1.36

Foods	Factor 3 uses		
	As a main dish	Eat with a fork	For dinner
Dip	1.07	1.29	1.69
Orange juice	1.09	1.01	1.84
Candy bars	1.09	1.15	1.26
Peanut butter	1.30	1.22	1.42
Dry cereal	1.64	1.12	1.26
Soft drinks	1.06	1.07	–
Ice cream	1.06	1.42	–
Wine	1.18	1.06	–
Apples	1.18	1.56	–
Coffee	1.20	1.05	–
Milk	1.20	1.10	–
Potato chips	1.20	1.16	–
Tea	1.26	1.04	–
Bread	1.26	1.38	–
Yogurt	1.87	1.58	–
Roast beef	6.74	6.51	6.74
Steak	6.70	6.78	6.88

Table 2.9 *Continued*

Foods	Factor 3 uses		
	As a main dish	Eat with a fork	For dinner
Spaghetti	6.44	6.67	6.53
Meat loaf	6.39	6.50	6.48
Fish	6.22	6.40	6.32
Ham	6.18	6.20	6.34
Chop suey	6.03	6.43	6.03
Chicken	6.57	–	6.60
Hamburger	6.16	–	6.16
Tossed salad	–	6.84	6.19

Foods	Factor 4 uses		
	At parties	With friends	For guests
TV dinners	1.12	1.76	1.22
Dry cereal	1.17	1.92	1.87
Chitterlings	1.48	1.71	1.68
Dip	6.54	6.13	6.48
Coffee	6.42	6.57	6.66
Wine	6.36	6.08	6.46
Cake	6.22	6.18	6.58
Shrimp	6.16	–	6.16

[a] Only consistent extreme means are reported, i.e., at least two out of the three factor-defining uses for each factor must be rated never appropriate (1.00 to 1.99) or always appropriate (6.00 to 6.99) for a given food.
[b] The three uses with the highest loadings on a factor are considered factor-defining uses.
(Adapted from Schutz *et al.*, 1975b.)

Table 2.10 Rank order of six highest appropriateness means for factor representative uses and preference for 58 foods

I 'Really like'	II 'For teenagers'	III 'Unhappy'
Steak	Ice cream	Coffee
Roast beef	Milk	Tea
Salad	Hamburger	Wine
Strawberries	Pizza	Ice cream
Spaghetti	Jello	Strawberries

IV 'As a main dish?'	V 'At parties'	VI 'Inexpensive'
Roast beef	Dip	Jello
Steak	Potato chips	Hamburger
Chicken	Coffee	Spaghetti
Spaghetti	Wine	Vegetable soup
Meat loaf	Cake	Meat loaf

(Adapted from Schutz *et al.*, 1975b.)

five foods for each use can be given with no means as illustrated in Table 2.10. Here the hedonic surrogate 'really like' is given plus the four highest weighted variables for each of the four use factors. The five highest appropriateness food means are ranked under each use. The data in this table are strong supportive evidence for the conclusion that the affective judgment of foods does not produce the same order as specific uses.

2.2.8 Analyses with non-appropriateness data

The frequencies for the demographic characteristics can be used to determine the success of representing a particular population, where such population figures are known and thus an estimate of the external validity of data. Characteristics such as age and income, if a sufficient sample size is available, can be used to produce separate subgroups for analysis. If data on frequency of consumption has been included it may be used as a dependent variable which can be predicted by principal components or surrogate representatives of each factor through multiple regression. This latter analysis can include, in addition, a separate measure of hedonic value for each food (if available) or a surrogate preference variable such as 'a food I really like' if included as one of the uses. Recently we have collected information on frequency of occasion (situation/use) as a possibly helpful variable in explaining consumption. Preliminary analyses, although they indicate wide differences in the frequency of occasion values for different uses, do not seem to add any predictive information with regard to frequency of consumption. Without any direct experimental evidence it appears that the construct of appropriateness has embedded within it some of the aspects of frequency of occasion, especially when information on appropriateness covers a wide range of uses.

Without appearing to overanalyze the data it is possible to select any one of the uses as a dependent variable in an attempt to predict it from other variables in the study such as other uses and the demographic characteristics. For example in a paper in preparation by Peter Rogers and the author the attribute 'difficult to resist' was selected as a measure of food craving and was successfully predicted for 50 foods as well as for individual foods.

I have described the analysis of foods and of uses using means over the population or subgroups of the population from the data. It is also possible to use individuals as the source of data for analyses. This was the case in the 'difficult to resist' study. In one earlier study (Rucker and Schutz, 1982) the classification of individuals was attempted utilizing a selected number of foods and uses from a much larger group. Although not earthshaking in its conclusions it did demonstrate that data on individuals could be utilized in this manner. In a presently ongoing study individual appropriateness data have also been used as raw data in preference mapping.

These are not the only techniques that could or should be used to derive useful statistical analysis of appropriateness data, but rather are the results of experience and examples of what can be done. Creative researchers may find other interesting ways in which the large volume and types of data collected in this manner might be analyzed.

2.3 Conclusions

In this chapter I have attempted to acquaint the reader with the rationale and procedures for utilizing the appropriateness technique as one way of looking at the cognitive-contextual aspects of food choice. It is certainly not a cookbook recipe for conducting such studies and there are obviously many unique concerns and considerations that will guide the researcher in the particular design of their studies. Some of the references that have been given where there are more details on procedures and results, may be helpful in the design of a new study. It is unquestionable that in order to improve our ability to understand and predict and perhaps even influence food choice we must have, in addition to measures of pure effect, some measure of the cognitive-context component of food. The appropriateness technique is one method to accomplish this objective but by no means is it the only way one can approach the question of cognitive-context characteristics. However in my opinion it is a rather simple but elegant procedure for collecting a large amount of relevant data in a relatively easy manner in a short period of time. Hopefully this chapter and the other methodologies presented in this book will result in sufficient familiarity and encouragement to result in the increased utilization of the cognitive-context component in a variety of food research.

References

Baird, P.R. and Schutz, H.G. (1976) The marketing concept applied to 'selling' good nutrition. *Journal of Nutrition Education*, **8**(1), 13–37.

Belk, R.W. (1975) Situational variables and consumer behavior. *Journal of Consumer Research*, **2**, December, p. 158.

Bruhn, C.M. and Schutz, H.G. (1986) Consumer perceptions of dairy and related-use foods. *Food Technology*, **40**(1), 79–85.

Gower, J.C. (1975) Generalised Procrustes analysis. *Psychometrika*, **40**, 33–51.

Kamenetzky, J., Pilgrim, F.J., and Schutz, H.G. (1957) Relationship of consumption to preference under different field conditions. Quartermaster Food and Container Interim Report, December.

Kakkar, P. and Lutz, R.J. (1981) Situational influence on consumer behavior: a review, in H.H. Kassarjiian and T.S. Robertson (eds) *Perspectives in Consumer Behavior*, 3rd edn, Scott Foresman, Glenview, IL, pp. 204–14.

Kelly, G.A. (1955) *The Psychology of Personal Constructs*, Vol. 1., Norton, New York.

Lau, D., Hanada, L., Kaminskyj, O., and Krondl, M. (1979) Predicting food use by measuring attitudes and preference. *Food Product Development*, **13**, 66–72.

Martens, M., Schutz, H.G., Risvik, E., and Rodbotten, M. (1988) Consumer perceptions of nutritional value related to other quality attributes and to chemical components, in J. Solms, D.R. Booth, R.M. Pangborn, and O. Raundhardt (eds) *Food Acceptance and Nutrition*, Academic Press, London.

McEwan, J.A. and Thomson, D.M.H. (1988) An investigation of the factors influencing consumer acceptance of chocolate confectionery using the repertory grid method, in D.M.H Thomson (ed) *Food Acceptability*, Elsevier Applied Science, London, pp. 347–62.

Miller, K.E. and Ginter, J.L. (1979) An investigation of situational variation in brand choice behavior and attitudes. *Journal of Marketing Research*, **16**, February, pp. 111–23.

Olson, J.C. (1981) The importance of cognitive processes and existing knowledge structures for understanding food acceptance, in J. Solms and R.C. Hall (eds) *Criteria of Food Acceptance*, Forster Verlag, Zurich, pp. 69–81.

Pilgrim, F.J. and Kamen, J.M. (1963) Predictors of human food consumption. *Science*, **139**, 501–2.

Rappaport, L. and Peters, G.R. (1988) Aging and the psychosocial problematics of food. *American Behavioral Scientist*, **32**, 31–40.

Resurreccion, A.V.A. (1986) Consumer use patterns for fresh and processed vegetable products. *Journal of Consumer Studies and Home Economics*, **10**, 317–32.

Rucker, M.H. and Schutz, H.G. (1982) Development of consumer typologies from appropriateness ratings. *Proceedings of the Academy of Marketing Science*, **5**, 587.

Rummel, R.J. (1970) *Applied Factor Analysis*. Northwestern University Press, Evanston, IL, p. 500.

Schutz, H.G. and Ortega, J.H. (1974) Consumer attitudes toward wine. *American Journal of Enology and Viticulture*, **25**, 33–8.

Schutz, H.G. and Rucker, M.H. (1975) A comparison of variable configurations across scale lengths: an empirical study. *Educational and Psychological Measurement*, **35**, 319–24.

Schutz, H.G., Rucker, M.H., and Hunt, J.D. (1971) Hospital patients' and employees' reactions to food-use combinations. *Journal of the American Dietetic Association*, **60**, 207–12.

Schutz, H.G., Fridgen, J.D., and Damrell, J.D. (1975a) Consumer perceptions of rice and related products. *Journal of Food Science*, **40**, 277–81.

Schutz, H.G., Rucker, M.H., and Russell, G.F. (1975b) Food and food use classification systems. *Food Technology*, **29**, 50–64.

Schutz, H.G., Moore, S.M., and Rucker, M.H. (1977) Predicting food purchase and use by multivariate attitudinal analysis. *Food Technology*, **31**, 85–92.

Schutz, H.G. (1988) Beyond preference: appropriateness as a measure of contextual evaluation of food, in D.M.H. Thomson (ed.) *Food Acceptability*, Elsevier Applied Science, London, pp. 115–34.

Sidel, J.L., Stone, H., Wollsey, A., and Macredy, J.M. (1972) Correlation between hedonic ratings and consumption of beer. *Journal of Food Science*, **37**, 335.

Stefflre, V. (1971) Some eliciting and computational procedures for descriptive semantics, in P. Kay, P. Reich, and M. McClaran (eds) *Explorations in Mathematical Anthropology*, MIT Press, Cambridge, MA, pp. 211–48.

Thomson, D.M.H. and McEwan, J.A. (1988) An application of the repertory grid method to investigate consumer perceptions of foods. *Appetite*, **10**, 181–93.

Wind, Y. (1977) Toward a change in the focus of marketing analysis: from a single brand to an assortment. *Journal of Marketing*, **41**, October, pp. 12, 143.

Worsley, A. (1980) Thought for food: investigations of cognitive aspects of food. *Ecology of Food and Nutrition*, **9**, 65–80.

3 The repertory grid approach
N. GAINS

3.1 Introduction

3.1.1 Food choice

How do we measure food preferences? This is the general title of the book, and each chapter attempts to address this question in a different way. Some might say that measurement of food preferences is intrinsically easy – people generally know what they like and dislike. However, understanding *why* consumers like one thing and dislike another is not so easy. Such understanding is vital in any 'real-life' commercial context; marketing, advertising, new product development, product positioning and product tracking all require that food manufacturers understand their market, their product's place in that market and the characteristics that define that product relative to others.

Many approaches used by market researchers such as pre-structured questionnaires, focus groups and free elicitation suffer from problems associated with limited response options, bias due to interviewers or dominant group members, and difficulty in describing product characteristics, particularly when products are viewed in isolation. The approach adopted in George Kelly's repertory grid procedures eliminates these problems. Allied with a recently developed multivariate analysis procedure, triadic elicitation as proposed by Kelly (1955) eliminates interviewer bias by allowing respondents to identify their own constructs and is sufficiently structured that the set of constructs they identify fully reflects the differences they perceive between the objects under investigation. Further, these terms, if subsequently used in attitude questionnaires, are likely to be more meaningful to consumers than terms generated by the researcher (Hughes, 1974).

This chapter describes this methodology for investigating consumer perceptions of foods, centred upon the repertory grid method, which is based on a sound psychological theory and offers the food researcher great flexibility.

3.1.2   *Personal construct theory*

Repertory grid method (RGM) is an integral part of George Kelly's theory of personal constructs (PCT; Kelly, 1955; Bannister and Fransella, 1986). According to this theory people act like scientists in the way they evaluate the world around them: formulating, testing, verifying and updating hypotheses about the world and its relationship to themselves (Gains, 1989). These hypotheses are realised as constructs, or bipolar dimensions, which describe two contrasting poles (e.g. bad–good, sweet–sour), which are hierarchically arranged into networks of constructs related to each other. In order to understand food preferences the researcher must investigate those constructs related to preference and choice, those constructs related to the consumer's perception of the foods and then relate the two together.

Repertory grid method is the term used to describe a set of techniques, related to Kelly's personal construct theory, which can be used to investigate the constructs individuals use to understand the world about them. There are many aspects to this world and indeed to food preference, but the method is flexible enough that it can be adapted to the researcher's interests. RGM as applied using triadic elicitation is described below.

3.2   Methodology

3.2.1   *Repertory grid method*

Kelly (1955) defines a construct as 'a way in which two things are alike and, in the same way, different from a third'. Consequently, triadic elicitation was developed for the elicitation of sets of constructs from individuals. Initially constructs related to an individual's perception of other people and the differences between them within the context of clinical psychology (Fransella and Bannister, 1977).

In practice the technique works as follows. Objects are arranged into groups of three (triads), such that each object appears in at least one triad and that one object from each triad is carried over to the next triad. Each triad is presented to the subject and two of the objects within that triad are arbitrarily associated with each other and dissociated from the third. The subject is then asked to describe how (s)he thinks that the two associated objects are similar and, in the same way, different from the third. As applied by the author, the procedure is extended to ask the subjects to describe the extremes of each elicited construct, so as to form a scale on which the objects can subsequently be quantified. Once the subject has exhausted all her/his constructs for that combination, the two remaining combinations within the triad are similarly presented to the subject and constructs elicited in the same way. This procedure is then repeated for each of the remaining triads, resulting in an exhaustive list of constructs (and associated extremes) which describe the terms in which that subject perceives the objects under investigation.

Constructs may then be associated with a 100 mm visual analogue scale (or any other appropriate scale), the poles of which are labelled with the extremes of the construct as described by the subject. Subjects may then use their individual lists of constructs, with corresponding scales, to rate each of the objects of interest. Thus, each subject defines a data matrix representing a particular multidimensional configuration of the objects.

3.2.2   Statistical analysis of repertory grids

Initially, repertory grid method was applied on an individual basis, normally within the context of clinical psychology. However, Kelly (1955) himself described how, for each individual, the internal cognitive relationships between objects, in the form of constructs, represent a form of multidimensional space. Indeed, he devised a form of non-parametric factor analysis which could be applied to repertory grids (Kelly, 1963). The idea of a mathematical space was considerably advanced by Slater (1964, 1977) who developed a range of programs, based on principal component analysis, for analysing repertory grids and producing perceptual spaces. Multidimensional scaling techniques have also been used in conjunction with RGM.

When investigating how subjects perceive a set of objects it is of great importance to be able to identify common dimensions of perception and experience across groups of these subjects. Slater (1977) used principal component analysis to study differences across repertory grids (RGs), but mainly in order to examine their reliability and consistency, and in most cases with either the same set of constructs or several common constructs. However, individuals tend to interpret construct descriptions in different ways and may also use scales differently. Slater's PREFAN program (Slater, 1977) forms a consensus interpretation by combining the RGs of a number of individuals into a single grid, but gives no information on individual differences that may be present, and results from such analyses may be biased toward certain individuals.

In any case, the idea behind the use of RGM is that individuals should be able to create their own unique set of constructs to describe a given set of objects. If there are common dimensions of perception across assessors, these will be manifest as geometrical similarities in the mathematical spaces defined by their repertory grids. That is, individuals may perceive the relationships between the objects as similar (in terms of structure), and may merely have defined their grids such that the orientations of these structures are different.

3.2.3   Generalised Procrustes analysis

Generalised Procrustes analysis (GPA; Gower, 1975) exploits this fact. It may be used to derive a perceptual map (configuration) of a set of objects from several sets of data or several repertory grids. The consensus configuration produced by GPA represents the main perceptual differences common to most or all of the grids. In addition, the technique allows the relative importance of each individual's constructs in defining the consensus configuration to be evaluated.

Procrustes analysis originated as a method for matching two multi-dimensional configurations (Hurley and Cattell, 1962). It works through the mathematical operations of transformation to a common origin, rotation/reflection of axes and dilation (i.e. uniform stretching and shrinking) to make one configuration approach the other as closely as possible (Gains *et al.*, 1988). The sum of the squared distances between the two configurations is known as the Procrustes statistic and is a measure of their closeness. GPA, an extension of the original method, works by iteratively matching each set of configurations to a consensus configuration, maximising the geometrical similarities between them (Gower, 1975). The method has been widely used in the sensory analysis of foods and beverages (Williams and Langron, 1984; Arnold and Williams, 1986) and is described in detail by Gains *et al.* (1988).

In the studies described below, GPA is applied to all the individual repertory grids (data matrices) to produce a consensus configuration and a matrix of distances between every individual data matrix. The consensus configuration is interpreted by relating (in this case correlating, although loadings or projections may also be used) all constructs to the dimensions of the configuration for each assessor in turn. Those constructs most highly correlated with each dimension for each assessor are then listed, and each dimension may then be interpreted on the basis of this list. The method will become apparent in the examples presented below.

3.3   Application

3.3.1   The different aspects of food choice

There are many aspects to food choice, and one of the advantages of repertory grid methodology as described in this chapter is that they can be adapted and applied to the investigation of all these aspects. Food choice is a psychological phenomenon with physical manifestations and

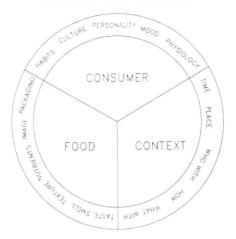

Figure 3.1 A schematic representation of the factors influencing food choice.

should be interpreted in the light of a theory of human behaviour such as Personal Construct Theory (Thomson and McEwan, 1988a; Gains, 1989). Generally, any form of food-related behaviour is the result of interaction between three things: the food itself, the consumer, and the context or situation within which this interaction takes place (Figure 3.1).

Food, consumer and context are themselves bundles of various factors and phenomena. For instance, food products have perceived sensory characteristics (which, of course, depend on the consumer), nutritional compositions, images, packagings and costs. Consumers have personalities, moods, physiological statuses, cultures, habits and memories which all affect their reactions to different foods. Finally, foods are not consumed in a void, but within specific contexts of use which greatly affect their acceptability. Context is a product of time, place, circumstance, manner and who and what the food is consumed with.

Thus, to understand the choice of any food product all these factors, if not necessarily measured, must be taken into consideration. For instance, it is important for a food manufacturer to understand *who* might buy their product, *what* they perceive the product to be like, and *where* and *when* they might consume it.

3.3.2 *Investigating the food: general perceptions, sensory characteristics and reasons for choice*

Only the most general application of the combined repertory grid/ generalised Procrustes analysis (RGM/GPA) approach has so far been described. This approach is exemplified by a recent case study of canned

Table 3.1 Canned lagers with original gravities and approximate price per 440 ml can (March 1988)

	Price	OG
Standard lagers		
Carling Black Label	49p	1034–38
Carlsberg	50p	1030–34
Castlemaine XXXX	56p	1033–37
Harp	47p	1030–34
Heineken	52p	1031–35
Hofmeister	42p	1034–38
Ind Coope Long Life	64p	1038–42
Kestrel	45p	1030–34
Miller Lite	67p	1030–34
Norseman	35p	1030–34
Skol	51p	1031–35
Premium lagers		
Grolsch	77p	1044–50
Kronenbourg 1664	70p	1043–49
Stella Artois	77p	1044–50
Pils lagers		
Holsten Pils	92p	1044–50
Super-strength lagers		
Carlsberg Special Brew	91p	1078–82
Tennents Super	92p	1082–86

lagers (Gains, 1989) in which consumers' perceptions of the lagers were investigated both before and after familiarisation with the products, as one part of the overall experimental strategy. Only the latter study will be reported here.

The consumers were 15 males and 5 females from the University of Reading and an industrial research centre on site. All were aged between 20 and 33 and regular drinkers of canned lager. Seventeen canned lagers, representing a range of type and price, were chosen from the products available in the south-east of the UK (Table 3.1).

Consumers were interrogated regarding their perceptions of the products using the normal triadic elicitation procedure and the question:

In what ways are these two products similar but, in the same way, different from the third?

Hence, each consumer supplied a list of constructs and anchor points, which were subsequently converted into scales such that each product could be rated for each construct. When data matrices had been standardised to the same size by the addition of 'dummy' zero variables (Gains *et al.*, 1988), they were analysed using GPA to produce a consensus configuration which was interpreted by correlating each individual's set of constructs with the dimensions of this configuration. Figures 3.2 and 3.3 show the first four principal axes of this configuration and Tables 3.2 to 3.5 list those constructs most highly correlated with the first, second, third and fourth principal axes for each consumer. The interpretations of the axes shown in Figures 3.2 and 3.3 are derived from these tables.

Along the first principal axis (PA1; Figure 3.2), Carlsberg Special Brew and Tennents Super are separated to the left, Holsten Pils, Stella Artois, Grolsch and Kronenbourg 1664 lie nearer the middle and all the other products lie on the right. This separation is due to alcohol content, price, mouthfeel and strength of flavour, with the three products on the left scoring higher for these attributes. Along the second axis (PA2; bottom to top of Figure 3.2) Norseman and Tennents Super, with Ind Coope Long Life and Kestrel to a lesser extent, are strongly associated with a UK image and origin, whilst Carlsberg, Heineken, Carling Black Label and Holsten Pils are considered to be freqently and better advertised, more familiar, more easily available and have a more foreign/continental image.

Along PA3 (Figure 3.3) consumers perceive Grolsch, Stella Artois and, to a lesser extent, Kronenbourg 1664 to have more eye-catching cans and a higher quality taste, whilst Carling Black Label and Castlemaine XXXX

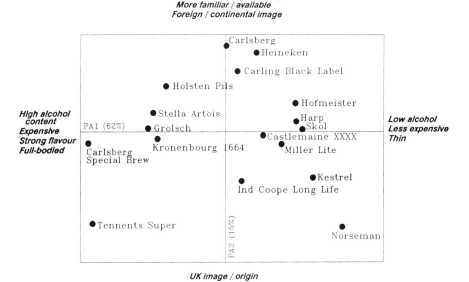

Figure 3.2 Canned lagers plotted in the plane defined by the first two principal axes of the consensus configuration (general perceptions study, Gains, 1989).

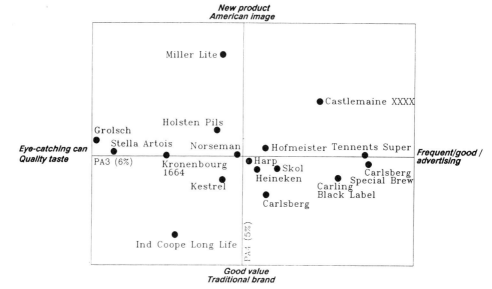

Figure 3.3 Canned lagers plotted in the plane defined by the third and fourth principal axes of the consensus configuration (general perceptions study, Gains, 1989).

Table 3.2 Constructs most highly correlated with the first principal axis for each subject (general perceptions study)

Consumer	Constructs	Respective correlations	
1.	Drink [alone–in company]; Alcoholic strength	0.94	−0.97
2.	Flavour strength; Body	−0.94	−0.92
3.	Strength (alcohol); Price	−0.93	−0.93
4.	Alcohol content; Cost	−0.93	−0.97
5.	Party; Association with name	−0.95	−0.97
6.	Aftertaste [short–long]; Bitterness	−0.88	−0.90
7.	Body; Price	−0.98	−0.95
8.	Strength; Light	−0.96	0.94
9.	'Naff'; Good preconceptions	0.97	−0.96
10.	Alcoholic strength; Cost	−0.79	−0.83
11.	Strength; Price	−0.88	−0.91
12.	Flavour strength; Price	−0.95	−0.95
13.	Fullness of flavour; Alcohol content	−0.97	−0.98
14.	Strength; Drink at home	−0.97	−0.95
15.	Alcohol strength; Body	−0.94	−0.96
16.	Price; Alcoholic strength	−0.94	−0.89
17.	Expense; Strength	−0.94	−0.95
18.	Strength; Price	−0.97	−0.97
19.	Alcohol strength; Fuller flavour	−0.97	−0.98
20.	Taste [full–weak]; Mouthfeel [thin–syrupy]	−0.99	−0.99

Table 3.3 Constructs most highly correlated with the second principal axis for each subject (general perceptions study)

Consumer	Constructs	Respective correlations	
1.	Give to friends; Standard offering	0.80	0.81
2.	Quality of advertising; Amount of advertising	0.79	0.70
3.	Character; Continental taste	0.83	0.81
4.	Extent of advertising; Availability	0.93	0.83
5.	Drink when 'out and about'; Advertisement (good)	0.84	0.83
6.	Drinkability; Bitterness	0.84	−0.84
7.	Familiarity; Frequency of advertising	0.89	0.79
8.	Aftertaste (nice); Smoothness	0.73	0.70
9.	Classy image; Foreign	0.83	0.80
10.	Home in evenings; Amount you can drink	0.90	0.90
11.	Origin [UK–abroad]; Marketing impact	0.60	0.80
12.	Can appeal; Head quality	0.63	0.74
13.	Pleasantness; Esteriness	0.58	−0.82
14.	Luchtime; Marketing awareness	0.73	0.87
15.	Lager flavour; Quality of advertising	0.82	0.82
16.	Eye-catching can; Watery	0.54	−0.46
17.	Brand names (known); Advertising (good)	0.74	0.85
18.	How well known		0.89
19.	Advertised; Continental image	0.85	0.78
20.	Advertising [dull–bright]; Continental image	0.89	0.49

Table 3.4 Constructs most highly correlated with the third principal axis for each subject (general perceptions study)

Consumer	Constructs	Respective correlations	
1.	Availability; Bitter	0.58	−0.71
2.	Viscosity; Value for money	0.55	−0.61
3.	Character; Continental taste	−0.95	−0.93
4.	Extent of advertising; Availability	0.63	0.91
5.	Bitter taste; Advertisement	0.42	0.45
6.	Advertising; Fizziness	0.75	−0.47
7.	Familiarity; Frequency of advertising	0.55	0.93
8.	Smoothness; Colour [light–dark]	−0.50	0.47
9.	Dull; Looks refreshing	0.92	−0.83
10.	Association with bottled lager; German image	−0.69	−0.60
11.	Can label design (attractive); Marketing impact	−0.50	0.56
12.	Can appeal; Head quality	−0.71	−0.75
13.	Metallic; Esteriness	0.67	0.53
14.	Taste; Marketing awareness	0.64	0.53
15.	Advertising profile; Green flavour	0.58	−0.62
16.	Body		0.50
17.	Refreshing; Advertising	−0.59	0.68
18.	Flavour [rough–smooth]; Quality of advertising	−0.75	−0.59
19.	Advertised; Macho image	0.72	0.77
20.	Advertising [dull–bright]		0.74

Table 3.5 Constructs most highly correlated with the fourth principal axis for each subject (general perceptions study)

Consumer	Constructs	Respective correlations	
1.	Value; Give to friends	−0.78	−0.62
2.	Trendiness; American image	0.63	0.73
3.	Character; Continental taste	−0.62	−0.85
4.	Newness; Refreshing	0.74	0.57
5.	Value for money; Drink with parents	−0.81	−0.81
6.	Advertising; Body	0.65	−0.73
7.	Familiarity		−0.57
8.	Aftertaste (nice); Fizziness	−0.63	0.62
9.	Dull; Eye-catching	−0.97	0.83
10.	American style beer		0.84
11.	Flavour; Brand tradition [old−new]	−0.74	−0.75
12.	Can style; Head quality	0.71	0.78
13.	Metallic		−0.44
14.	Taste; Drink to refresh	−0.52	−0.64
15.	Bitter; Novel product	−0.86	0.64
16.	Alcoholic strength; Watery	−0.73	0.67
17.	Brand names (known); Advertising	0.51	0.68
18.	Flavour [rough−smooth]; Design of can	−0.68	−0.74
19.	Brash; American image	0.91	0.85
20.	Price; Quality	0.64	0.65

in particular (right-hand side) are again described as well/frequently advertised. At the bottom of Figure 3.3 (PA4) Ind Coope Long Life is described as a traditional brand and good value for money, whilst at the top of this axis Miller Lite is considered a newer product with an American image.

This is an example of how the method may be used to investigate consumers' general perceptions of foods. The researcher may sometimes, however, wish to investigate a specific type of perception (e.g. how the consumer perceives the sensory characteristics of the products). By merely altering the question to one which is more appropriate, RGM may be applied as described above. For instance, if the researcher was interested in those differences between foods which specifically affect consumer choice then an appropriate question might be:

For what reasons would you buy/consume these two products but not the third?

Such a question does work very effectively by forcing the consumer to attend to those factors which actually influence her/his choice behaviour, although no such studies have yet been published. McEwan and Thomson (1988a) have published a comparative study in which triadic elicitation is one of the techniques used to investigate the sensory characteristics of chocolate.

3.3.3 Investigating consumers

When investigating food choice it is very important to be able to recognise differences across consumers when they do occur. Although the technique of GPA specifically recovers dimensions of common perception, these dimensions may not be common to every single consumer. The Procrustes statistic (a measure of the distance between two configurations) can be used to explore differences between all the configurations input into a GPA and also to explore the differences between these configurations and the output consensus configuration. Measures of fit between each dimension of this consensus configuration and the dimensions of each individual's configuration may also be obtained. Thus, it is possible to measure the 'consensusness' of each dimension and to investigate the nature of those consumers whose perceptions are represented by specific dimensions.

For instance, in the example of an investigation of general perceptions of canned lagers presented in the previous section, the Procrustes statistics (distances) from the GPA program may be input into a principal co-ordinate analysis (PCO; Chatfield and Collins, 1980) to produce a 'map' which represents the differences between consumers, highlighting groups

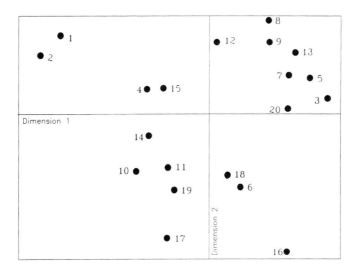

Figure 3.4 Plot of assessor differences as derived by principal co-ordinate analysis of the Procrustes distances (general perceptions study, Gains, 1989).

(or segments) of consumers with differing perceptions. Such a map is shown as Figure 3.4. There are two major groups in this plot and several outliers, although it is difficult to investigate these in detail because of the relatively small number of consumers used. Further investigation of these groups revealed that the differences were not significant and did not invalidate the integrity of the consensus configuration that was recovered. However, in some cases such differences may be important.

Table 3.6 gives the percentage variance accounted for by projecting the principal components of each assessor's data matrix (after transformations) onto the principal components of the consensus configuration for the same study. This provides a measure of the 'importance' each consumer attaches to each dimension of this configuration (i.e. a measure of fit), and also allows the researcher to assess the unanimity (or consensusness) of the perception(s) represented by each of these dimensions. In this case it appears that there is a strong degree of consensus with regard to the major dimensions and that the consensus configuration is a fair representation of the perceptions of nearly all of the consumers, with one or two exceptions.

Further evidence of the extent to which the perceptions of the con-

Table 3.6 Percentage variance accounted for by projecting on to first five principal axes of consensus configuration for each assessor (general perceptions study)

	Principal component				
Assessor	1	2	3	4	5
1	49	15	5	7	7
2	45	13	7	6	7
3	47	13	24	5	6
4	46	20	7	11	4
5	61	11	3	4	6
6	31	32	17	7	3
7	63	22	5	4	2
8	66	7	4	7	6
9	67	11	11	6	2
10	25	31	7	8	8
11	29	17	7	16	18
12	68	9	5	3	5
13	76	10	2	2	3
14	48	18	8	9	7
15	50	19	5	5	2
16	54	4	6	9	5
17	28	33	14	8	7
18	60	13	6	2	3
19	53	18	8	8	3
20	82	10	3	0	2
Consensus	62	15	6	5	4

sumers correspond can be garnered from Tables 3.2 to 3.5 which list those constructs most highly correlated with the first four dimensions of the consensus configuration. If the individual consumer's constructs best fitted to each dimension are similar, then it may be concluded that they perceive the products in much the same way. Of course, this is in fact the basis on which the dimensions are interpreted. These tables also highlight where consumers perceive the similarities and differences between the products in much the same way, but use different words to describe these differences.

This is not the only way in which individual differences may be investigated. Differences across consumers may also be investigated directly using RGM. For instance, it is possible to study the 'sort' of people that consumers associate with particular products (i.e. stereotypes). Such information is very important in assessing a marketing and advertising strategy for the product and the target population to whom it should be directed. A suitable question, in the context of triadic elicitation, might be:

> What sort of people do you think might use these two products but not the third?

This is one of the many bases on which consumers classify products, and their perception of the suitability of products for different types of people may directly affect their choice behaviour.

3.3.4 Investigating contexts of use

Context of use is a very important factor in food choice (Schutz, 1988; Scriven et al., 1989). For instance, the difference in consumption between black and white coffee may often be put down to context of use. Scriven et al. (1989) investigated the contexts of use of a range of alcoholic beverages using the repertory grid method, and in many ways the work reported here is an extension of this study to a much tighter product range of canned lagers.

Contexts of use differentiating the products were elicited using triadic elicitation as described previously, and the question:

> In what contexts would you consume these two products but not the third?

Context of use was operationally defined as a time, manner, place or circumstance in which a food or beverage might be consumed, and this idea was explained to consumers (in most cases unnecessarily). Once consumers had formed a list of contexts, each context was associated with a 100 mm visual analogue scale labelled 'never' and 'always' at the left and right hand extremities respectively. Consumers were asked to rate the 'appropriateness' of each canned lager for use in each of their own

Table 3.7 Constructs most highly correlated with the first principal axis for each subject (context study)

Consumer	Constructs	Respective correlations	
1.	In pub, when draught poor; Treat	−0.96	−0.95
2.	For lager drinker; For own indulgence	−0.94	−0.93
3.	Get drunk; With dinner	−0.92	0.64
4.	Thirst quencher; Impress people	0.89	−0.89
5.	To get drunk; Celebration	−0.98	−0.96
6.	Get drunk quickly; Not much money	−0.89	0.89
7.	Slava; One or two friends	0.95	−0.75
8.	To get drunk; Going to football	−0.94	−0.90
9.	Party; To enjoy	−0.96	−0.95
10.	Barbecues; Picnics	0.95	0.94
11.	Driving; Sporting activities	0.96	0.94
12.	Driving; Cash flow problem	0.95	0.89
13.	At home; Take out	−0.94	−0.90
14.	Evening; With company	−0.93	−0.89
15.	After sport; For refreshment	0.99	0.97
16.	Eating takeaway; Party	0.58	−0.40
17.	Party; Picnic	1.00	0.97
18.	To get slightly drunk; Thirsty	−0.86	0.74
19.	Watching TV; With snacks	−1.00	−1.00
20.	Football watching; Party	−0.98	−0.98

Table 3.8 Constructs most highly correlated with the second principal axis for each subject (context study)

Consumer	Constructs	Respective correlations	
1.	For refreshment; 1 or 2, but want to be satiated	0.80	0.81
2.	To take to a party; To drink at a party	0.81	0.80
3.	Watching TV/At home; At a party	0.96	0.96
4.	With cashflow problem		0.64
5.	With a meal (takeaway); As a thirst quencher	0.91	0.91
6.	Hot day; Not much money	0.76	0.73
7.	On a trip: Friends come round	0.97	0.91
8.	Down pub; When playing cards	0.86	0.85
9.	Summer afternoon; Outing	0.90	0.77
10.	In groups; Watching TV	0.94	0.87
11.	At friends; Weekends/Holidays	0.84	0.79
12.	In a pub: After sport	0.96	0.90
13.	Sitting in garden; Parties	0.95	0.94
14.	With relatives: After sport	0.65	0.54
15.	Before meal; At home	0.96	0.95
16.	Watching telly; At home	0.96	0.96
17.	Christmas; Home	0.97	0.97
18.	Summer (outside); Initial	0.83	0.81
19.	With meals; Drink away from home	0.99	0.99
20.	Thirst quenching; Watching TV	0.97	0.86

Table 3.9 Constructs most highly correlated with the third principal axis for each subject (context study)

Consumer	Constructs	Respective correlations	
1.	Quantity and quality; For dedicated lager drinker	0.81	0.74
2.	With a snack; With groups of people	0.74	0.69
3.	Wine bar: In pub	0.67	0.51
4.	Quiet drink; Under stress	0.54	−0.50
5.	Camping/hiking (outdoor pursuits); Party	−0.75	−0.55
6.	On the train		−0.66
7.	On a trip; Friends come round	−0.94	−0.88
8.	At home; Going down park	−0.75	−0.74
9.	Summer afternoon; Home	0.90	0.87
10.	Alone; In groups	−0.67	−0.55
11.	Summer outside; At home	−0.71	−0.70
12.	After meal; In a pub	0.98	0.92
13.	Friends' homes		0.43
14.	With relatives; At home	−0.75	−0.63
15.	Hot weather; Hiking	−0.94	−0.93
16.	In the garden; Eating takeaway	−0.72	−0.58
17.	Visitors; Party	−0.80	−0.68
18.	Weekends; Summer (outside)	0.87	0.77
19.	Drink away from home; At party	0.94	0.92
20.	Thirst quenching; Ethnic occasion	0.66	−0.56

Table 3.10 Percentage variance accounted for by projecting on to first five principal axes of consensus configuration for each assessor (context study)

Assessor	Principal component				
	1	2	3	4	5
1	67	3	12	7	3
2	55	12	7	3	5
3	35	47	8	4	1
4	65	4	5	3	4
5	40	32	7	7	3
6	66	14	7	4	1
7	45	45	5	2	1
8	38	34	5	4	3
9	68	7	11	3	1
10	47	31	5	4	3
11	50	10	12	3	18
12	48	29	15	2	1
13	27	52	2	8	4
14	50	6	13	11	8
15	38	43	4	3	7
16	10	76	5	2	0
17	33	45	3	2	2
18	37	31	10	2	5
19	68	22	5	3	1
20	71	12	5	1	4
Consensus	60	21	7	3	3

contexts of use on this scale. Data matrices were standardised in size and then analysed using GPA.

Results are presented as tables and figures. Those constructs most highly correlated with each dimension of the consensus configuration are listed in Tables 3.7 to 3.9 (the solution was three dimensional – see Table 3.10). These tables were used to interpret Figures 3.5 and 3.6 which show the first against the second, and the first against the third principal axes of this consensus configuration.

Along the first dimension (Figure 3.5 – left to right) the canned lagers are separated as in the previous study of general perceptions, with the super-strength, premium and Pils lagers on the left-hand side of the plot and the standard lagers on the right. The standard lagers are considered more appropriate with meals and for outdoor activities. The premium and super-strength lagers are described as more appropriate for a treat or indulgence, to get drunk, and on special occasions.

Along the second dimension (Figure 3.5 – bottom to top), those products at the top of the figure (Stella Artois, Kronenbourg 1664, Holsten Pils, Carlsberg, Heineken, Hofmeister, Harp and Carling Black Label) are considered more appropriate to drink at home and also at parties and away from home. The fact that the contexts 'At home' and 'Away from home' both lie in the same direction implies nothing more than that those lagers at the top of the figure are considered more appropriate in these situations: such contexts are not mutually exclusive.

In the top right quadrant of Figure 3.5, Carlsberg, Heineken,

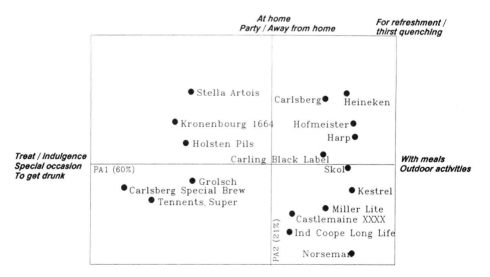

Figure 3.5 Canned lagers plotted in the plane defined by the first two principal axes of the consensus configuration (contexts study; source: Gains and Thomson, 1990a, p. 702).

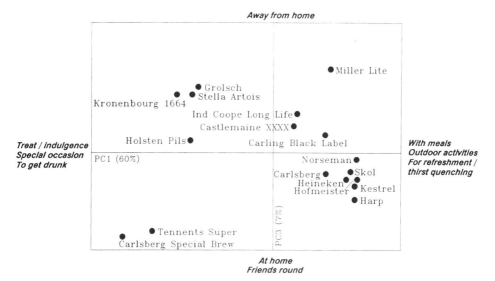

Figure 3.6 Canned lagers plotted in the plane defined by the first and third principal axes of the consensus configuration (contexts study; source: Gains and Thomson, 1990a, p. 702).

Hofmeister, Harp and Carling Black Label are strongly associated with contexts involving refreshment and quenching of thirst.

Along the third dimension (Figure 3.6 – bottom to top) there is separation between those lagers drunk at home and those drunk away from home. Specifically, the two super-strength lagers (Carlsberg Special Brew and Tennents Super) are more generally drunk at home or with friends round, whilst Grolsch, Stella Artois, Kronenbourg 1664 (the three premium lagers) and Miller Lite are drunk away from home. As previously, readers are referred to Tables 3.7 to 3.9 for complete interpretations of all these dimensions.

3.3.5   Interaction of foods, consumers and context of use

It is apparent that there are many similarities between the configuration derived in the general perceptions study of the lagers and the configuration derived from differences in context of use as described above. Thus the researcher can begin to piece together the characteristics of products which make them appropriate within specific contexts of use. It is true that consumers choose bundles of qualities rather than specific food items *per se* (as George Berkeley stated in 1710). A consumer's needs are characterised by her/his own requirements and the context within which the interaction between food and consumer takes place. Foods (merely?) supply characteristics which fulfil these needs to a lesser or greater extent.

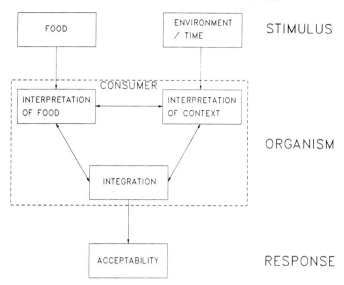

Figure 3.7 The interaction between consumer, food and context to form a food acceptance response (behaviour).

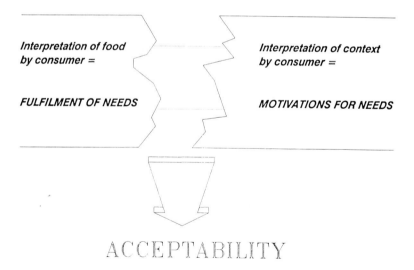

Figure 3.8 The interaction between a consumer's perceptions of the food and the context of use.

Figures 3.7 and 3.8 depict this process as described by Gains (1989), and are the basis of his model of food choice.

Thus it is quite possible to investigate perceptions of foods without referring to specific products or brands which may, in any case, focus the consumer's attention toward specific characteristics of the products. Gains (1989) used RGM to investigate perceptions of foods in this way, as part of a wider investigation of the factors affecting consumer choice of canned lagers.

Briefly, a consensus list of the eighteen most frequently described contexts was compiled from those elicited in the contextual study already described. These contexts were: 'Down the pub', 'With relatives', 'With friends', 'Outing/Picnic', 'Short of money', 'Summer afternoon', 'At home', 'Party', 'To get drunk', 'Celebration/Special occasion', 'Watching sport', 'With a meal/take-away', 'For refreshment', 'Watching TV', 'Weekend', 'Driving', 'After sport' and 'Travelling'. These contexts were written on cards and an interview conducted with each assessor to elicit the differences between the contexts, in terms of the characteristics of canned lagers considered most relevant to these contexts. On presentation of the triads, assessors were asked:

> What attributes would you consider important in a canned lager in these two contexts, which you would consider less important in the third?

Thus, a list of characteristics, with appropriate poles, was elicited from each assessor. Each characteristic was subsequently associated with a 100 mm visual analogue scale and assessors rated the ideal level of each characteristic for a canned lager within each of the eighteen contexts of use.

The data were analysed using GPA, to produce a map of the contexts constructed from the perceived differences in appropriate characteristics of canned lagers, which consisted of four interpretable dimensions (Tables 3.11 to 3.15). The configuration was interpreted by correlating the assessor's original constructs with the dimensions of the configuration as described above.

Along the first dimension (Figure 3.9 – left to right) the contexts are separated by characteristics related to alcohol content. To the left of the plot the contexts 'To get drunk' and, to a lesser extent, 'Celebration/ Special occasion' are associated with stronger and heavier lagers, whilst to the right lagers are considered more appropriate for 'Driving', 'Refreshment', 'After sport' and 'Summer afternoons' if they are lighter and weaker. Particularly in the bottom right corner, the contexts 'For refreshment', 'After sport' and 'Summer afternoon' are associated with refreshing products.

From bottom to top along the second dimension (Figure 3.9), the contexts 'Celebration/Special occasion', 'For refreshment', 'Summer

Table 3.11 Constructs most highly correlated with the first principal axis for each subject (second context study)

Consumer	Constructs	Respective correlations	
1.	Lightness [light–heavy]; Strength (COOH)	−0.90	−0.85
2.	Alcohol content; Lightness [light–heavy]	−0.96	−0.77
3.	Alcoholic strength; Availability	−0.95	−0.50
4.	Alcoholic strength; Refreshability	−0.78	0.72
5.	Bitterness; Strength (alcohol)	−0.91	−0.90
6.	Strength; Price	−0.85	−0.82
7.	Alcohol content; Body	−0.99	−0.97
8.	Strength; Flavour strength	−0.96	−0.94
9.	Alcohol level; Refreshing	−0.88	0.84
10.	Alcohol; Strength	−0.92	−0.92
11.	Thirst quenching; Price	0.85	−0.79
12.	Colour [light–dark]; Appeal	−0.94	−0.92
13.	Flavour strength; Alcoholic strength	−0.72	−0.71
14.	Alcoholic strength; Price	−0.90	−0.68
15.	Alcohol content; Price	−0.89	−0.89
16.	Alcohol; Body	−0.86	−0.84
17.	With lemonade; Expensive	−0.92	−0.85
18.	Cost; Intensity of flavour	−0.89	−0.87
19.	Quality; Alcohol strength	−0.99	−0.98
20.	Mouthfeel/Body; Alcohol content	−0.92	−0.91

Table 3.12 Constructs most highly correlated with the second principal axis for each subject (second context study)

Consumer	Constructs	Respective correlations	
1.	Value; Thirst quenching	0.66	−0.62
2.	Clean taste; Fizz potential	−0.75	−0.65
3.	Price; Availability	−0.74	−0.73
4.	Refreshability; Cost	−0.64	−0.57
5.	Satisfying; Taste quality	−0.82	−0.80
6.	Effervescence; Refreshing	−0.74	−0.64
7.	Taste [smooth–bitter]; Price	0.89	−0.81
8.	Taste; Fizziness	−0.67	−0.63
9.	Fizziness; Quality	−0.75	−0.55
10.	Cost image; Cost	−0.72	−0.71
11.	Brand image; Flavour quality	−0.68	−0.64
12.	Price; Refreshing	−0.88	−0.81
13.	Refreshing; Sweetness	−0.73	−0.53
14.	Fashionable [old–foreign/chic]; Price	−0.84	−0.78
15.	Carbonation; Thirst quenching	−0.55	−0.52
16.	Price; Choice [none–lots]	−0.88	−0.70
17.	Strong; Expensive	−0.67	−0.54
18.	Cost; Smoothness	−0.55	−0.50
19.	Body; Availability	−0.91	−0.89
20.	Carbonation; Quality image	−0.48	−0.48

Table 3.13 Constructs most highly correlated with the third principal axis for each subject (second context study)

Consumer	Constructs	Respective correlations	
1.	Image [bottom–top]; Share a pack with others	−0.61	−0.49
2.	Popularity; Lightness	−0.89	−0.60
3.	Availability		0.85
4.	Alcoholic strengh; Strength of flavour	0.66	−0.63
5.	Taste (quality); Strength (alcohol)	0.55	0.51
6.	Price		0.55
7.	Flavour		0.42
8.	Taste		−0.81
9.	Drinkability; Innocuous	0.84	−0.52
10.	Sweetness; Size of can	0.67	0.63
11.	Gassy		0.64
12.	Perceived quality; Mellow [sharp–mellow]	−0.66	−0.64
13.	Price; Bitterness	0.82	0.81
14.	Fizziness		0.78
15.	Overall quality; Thirst quenching	−0.65	0.55
16.	Quality; Price	−0.89	−0.78
17.	Gassy; Strong	−0.90	0.65
18.	Design of can (distinct)		−0.75
19.	Availability; Alcohol strength	0.92	0.87
20.	Sweetness; Carbonation	0.70	0.46

Table 3.14 Constructs most highly correlated with the fourth principal axis for each subject (second context study)

Consumer	Constructs	Respective correlations	
1.	Share a pack with others; Gassiness	0.68	0.64
2.	Value for money (bulk); Taste (quality)	0.89	0.52
3.	Taste quality; Price	−0.91	−0.70
4.	Strength of flavour		−0.74
5.	Value for money		0.76
6.	Brand popularity; Price	−0.72	−0.71
7.	Taste [smooth–bitter]		−0.79
8.	Fizziness		0.79
9.	Interesting		0.65
10.	Fizziness; Bitterness	0.69	0.62
11.	Gassy		0.84
12.	Refreshing		0.61
13.	Alcoholic strength; Acidity	0.69	0.65
14.	Fizziness; Alcoholic strength	0.47	0.44
15.	Product familiarity; Bitter	0.77	0.62
16.	Fizziness; Quality	0.91	0.80
17.	Gassy; Expensive	0.85	−0.70
18.	Fizzziness		0.95
19.	Availability; Flavour	0.80	0.56
20.	Full flavour; Alcohol content	0.69	0.55

Table 3.15 Percentage variance accounted for by projecting on to first five principal axes of consensus configuration for each assessor (second context study)

	Principal component				
Assessor	1	2	3	4	5
1	39	21	13	4	6
2	39	13	12	19	5
3	33	24	16	1	13
4	43	17	17	6	11
5	50	17	7	8	4
6	60	21	4	2	2
7	84	8	2	2	1
8	69	10	7	5	1
9	37	14	15	4	12
10	32	22	18	10	4
11	54	17	7	8	5
12	54	22	16	1	2
13	30	12	33	11	3
14	41	34	10	5	4
15	38	12	6	8	10
16	48	17	9	7	10
17	58	5	11	9	9
18	58	8	7	12	4
19	75	10	6	2	5
20	61	10	7	9	6
Consensus	60	15	10	5	4

afternoon' and 'After sport', in particular, are associated with fizzy and satisfying lagers. The context 'Celebration/Special occasion' is also associated with expensive canned lagers, whilst cheaper lagers are considered more appropriate when 'Short of money'.

Along the third dimension (left to right – Figure 3.10) the contexts 'With friends', 'With relatives' and 'Driving' are differentiated by association with high quality canned lagers whilst 'For refreshment', 'After sport' and 'To get drunk' are differentiated by association with strong and refreshing lagers, although to some extent this merely reflects a polarisation with the opposite side of the plot. Going from bottom to top in Figure 3.10 (fourth dimension), expensive products are considered more appropriate 'To get drunk', 'Travelling' and 'Driving' whilst value for money is considered more important in the contexts 'Short of money', 'Summer afternoon', 'At home' and 'Outing/picnic'.

Context is an important factor in food choice and this study represents one approach to its investigation. In particular, it allows the investigator to study the nature of products appropriate within different contexts without reference to branded products which sometimes lead consumers

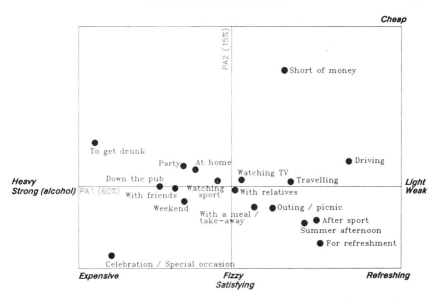

Figure 3.9 Plot of canned lager drinking contexts in the plane defined by the first two principal axes of the derived consensus configuration (second context study).

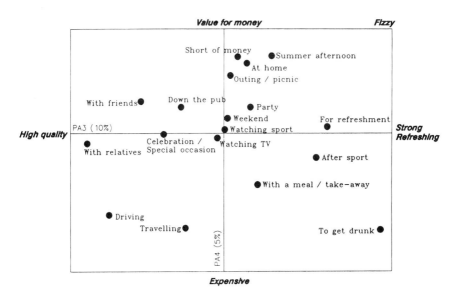

Figure 3.10 Plot of canned lager drinking contexts in the plane defined by the third and fourth principal axes of the derived consensus configuration (second context study).

to concentrate on the more visual aspects of products (e.g. branding, packaging and style).

3.3.6   Extending repertory grid methodology: laddering, preference mapping and other procedures

Repertory grid methodology offers a complete and thorough system for the qualitative and quantitative assessment of all the perceptions employed by consumers to differentiate products and product types. Ultimately, the researcher will need to relate these perceptions to preferences for the different products, and many multivariate techniques (e.g. preference mapping) may be used for this purpose as they are with other types of data (McEwan and Thomson, 1989).

A market researcher may also wish to investigate the motivations that drive individual consumers in even greater depth than may be achieved through RGM. For instance, they may seek to understand the structure of the relationships between the perceptions an individual holds or to uncover some of the more hedonic and personal motivations underlying food choice. One technique in particular has been developed to such a purpose, again within the context of clinical psychology and personal construct theory. This is *laddering* as developed by Hinkle (1965). In essence, this technique works by tracing a path through an individual's construct hierarchies. Landfield (1971) has developed a related 'pyramid' procedure.

3.4   Conclusions

Repertory grid method (RGM) provides a technology sufficiently flexible that it may be adapted to the needs of each study, but sufficiently structured that it allows consumers to fully verbalise their perceptions of the products under investigation. Allied with generalised Procrustes analysis (GPA) and other multivariate statistical procedures it offers a reliable route to understanding how consumers perceive any range of products – in terms of the sensory characteristics, general characteristics, image and packaging of the products themselves, in terms of the contexts within which different consumers use the products, and in terms of who might use the product. It is vitally important that all these aspects of existing and newly developed products are understood by the marketer, and once they are, all the ingredients in the marketing mix will fall into place. The application of the combined repertory grid method/ generalised Procrustes analysis approach in many studies attests to its fundamental efficacy (McEwan and Thomson, 1988b, 1989; Thomson and

McEwan, 1988; Scriven *et al.*, 1989; Gains and Thomson, 1990a, 1990b) and it deserves to be more widely applied.

References

Arnold, G.M. and Williams, A.A. (1986) The use of generalised Procrustes techniques in sensory analysis, in J.R. Piggott, (ed.) *Statistical Procedures in Food Research*, Elsevier Applied Science, London, pp. 233–53.

Bannister, D. and Fransella, F. (1986) *Inquiring Man: The Psychology of Personal Constructs*, Croom Helm, New York.

Berkeley, G. (1710) *An Essay Towards a New Theory of Vision*, Dublin.

Chatfield, C. and Collins, A.J. (1980) *Introduction to Multivariate Analysis*, Chapman and Hall, London.

Fransella, F. and Bannister, D. (1977) *A Manual for Repertory Grid Technique*, Academic Press, London.

Gains, N. (1989) The integration of personal construct theory in food acceptability research, PhD thesis, University of Reading, UK.

Gains, N. and Thomson, D.M.H. (1990a) Contextual evaluation of canned lagers using repertory grid method. *International Journal of Food Science and Technology*, **25**, 699–705.

Gains, N. and Thomson, D.M.H. (1990b) The relation of repertory grid generalised Procrustes analysis solutions to the dimensions of perception: Application to Munsell color stimuli. *Journal of Sensory Studies*, **5**, 177–92.

Gains, N., Krzanowski, W.J., and Thomson, D.M.H. (1988) A comparison of variable reduction techniques in an attitudinal investigation of meat products. *Journal of Sensory Studies*, **3**, 37–48.

Gower, J.C. (1975) Generalized Procrustes analysis. *Psychometrika*, **40**, 33–51.

Hinkle, D. (1965) The change of personal constructs from the viewpoint of construct implications, PhD thesis, Ohio State University, USA.

Hughes, G.D. (1974) The measurement of beliefs and attitudes, in R. Ferber, (ed.) *Handbook of Marketing Research*, Appleton-Century-Crofts, New York, pp. 316–43.

Hurley, J.R. and Cattell, R.B. (1962) The Procrustes program: producing direct rotation to test a hypothesised factor structure. *Behavioural Science*, **7**, 258–62.

Kelly, G.A. (1955) *The Psychology of Personal Constructs*. WW & Norton Co., New York.

Kelly, G.A. (1963) Nonparametric factor analysis of personality theories. *Journal of Individual Psychology*, **19**, 115–47.

Landfield, A.W. (1971) *Personal Construct Systems in Psychotherapy*, Rand McNally, New York.

McEwan, J.A. and Thomson, D.M.H. (1988a) A behavioural interpretation of food acceptability. *Food Quality and Preference*, **1**, 4–11.

McEwan, J.A. and Thomson, D.M.H. (1988b) An investigation of the factors influencing consumer acceptance of chocolate confectionery using the repertory grid method, in D.M.H. Thomson, (ed.) *Food Acceptability*, Elsevier Applied Science, London, pp. 347–62.

McEwan, J.A. and Thomson, D.M.H. (1989) The repertory grid method and preference mapping in market research: a case study on chocolate confectionery. *Food Quality and Preference*, **1**(2), 59–68.

Schutz, H.G. (1988) Beyond preference: Appropriateness as a measure of contextual evaluation of food, in D.M.H. Thomson (ed.), *Food Acceptability*, Elsevier Applied Science, London, pp. 115–34.

Scriven, F.M., Gains, N., Green, S.R., and Thomson, D.M.H. (1989) A contextual evaluation of alcoholic beverages using the repertory grid method. *International Journal of Food Science and Technology*, **24**, 173–82.

Slater, P. (1964) *The Principal Components of a Repertory Grid*, Vincent Andrews, London.

Slater, P. (1977) *The Measurement of Intrapersonal Space By Grid Technique*, Vols 1 and 2, Wiley, Chichester.

Thomson, D.M.H. and McEwan, J.A. (1988) An application of the repertory grid method to investigate consumer perceptions of foods. *Appetite*, **10**, 181–93.

Williams, A.A. and Langron, S.P. (1984) Use of free-choice profiling for evaluation of commercial ports. *Journal of the Science of Food and Agriculture*, **35**, 558–68.

4 Focus group interviewing

M.A. CASEY and R.A. KRUEGER

4.1 Introduction

Every methodology in social science research includes a set of recommended procedures intended to minimize abuse and error. Because these
standards are still evolving for focus groups, the methodology is prone to
abuses, both intentional and unintentional. For example, many disparate
group situations are labeled 'focus group interview'; researchers are
occasionally more interested in profit than in enlightenment; and results
are frequently overgeneralized. In this chapter we hope to highlight those
factors which are becoming accepted as critical to successful focus group
interviews and also those 'black holes' that weaken the quality of focus
group research.

A focus group interview is a carefully planned session designed to
obtain several individuals' perceptions of a defined area of interest
in a permissive, non-threatening environment. It is conducted with
approximately six to nine people by a skilled moderator. (We prefer the
term moderator to interviewer because it more accurately describes the
person's function.) The discussion is relaxed, comfortable, and often
enjoyable for participants. Careful and systematic analysis of the discussions provides insights into how a product, service, or opportunity is
perceived (Krueger, 1994).

4.1.1 Advantages and limitations of focus groups

As a research method, focus group interviewing offers several advantages.
First, and most important, it encourages and captures interaction among
participants. This interaction provides the researcher with a chance to
observe how people influence one another and how they talk about
specific topics. The technique provides information that is difficult to
obtain with other methods. Second, the focus group format allows the
moderator to probe for additional information at critical points. Third,
focus group discussions have high face validity. Clients can easily understand the technique and typically find the results credible. Fourth, the
method enables researchers to increase the sample size of qualitative
studies without dramatically increasing the time required to conduct them.

Fifth, an organization conducting a focus group is looked upon favorably by participants. They are impressed by the organization's concern about consumers' opinions and its willingness to listen. Participants often appreciate the chance to provide input.

There are also limitations to focus groups. First, moderators must be trained to conduct the interviews. The moderator must become skillful in encouraging the participation of everyone and preventing a few individuals from dominating the group's interactions. Moderators must be masterful at eliciting information from others while holding in check their personal points of view. Second, the data are difficult to analyze. The group interaction offers rich dynamics but also makes interpretation difficult. And analysis requires time and thought. Third, a number of groups should be conducted because groups vary considerably. One group may be so unresponsive that it takes all the moderator's skill to get any discussion going. The next may be so loquacious that the moderator needs to keep them on track. At least three groups should be conducted to balance idiosyncrasies among groups. Fourth, groups are difficult to assemble. Identification and recruitment of participants is time-consuming and requires systematic processes with sufficient incentives.

4.1.2   Myths about focus groups

Over the past decade a number of myths have developed about focus group research, among them the following:

Myth: Focus groups are inexpensive.

Fact: The cost of conducting focus groups varies greatly. Inexpensive is a relative concept. If you are accustomed to spending $50 000 for a mail-out survey, then a $10 000 focus group study is inexpensive. Nonprofits can keep costs low by using volunteers (moderators, participants, analysts), borrowing equipment (microphones, tape recorders) and paying only for refreshments (which are sometimes donated). Focus group rates quoted in trade journals average about $4000 per group, so that a typical study with three or four groups costs $12 000 to $16 000.

Myth: Focus groups require one-way mirrors.

Fact: Experts are split on the benefit and value of on-site analysis by observers versus analysis and synthesis by the moderator. The primary benefit of the one-way mirror is that it allows a team of researchers to see and hear the discussion. If the research team has previously had limited contact with consumers, there might be considerable benefit to having them attend the focus group. This then obligates the research team to attend each of the focus groups to avoid missing critical data, thus greatly increasing the cost of the focus group study. On the other hand, a competent, trusted analyst may be able to accomplish the same (or better) results because with the same financial investment the number of groups can be increased, the environment can be more natural (no one-way mirror), and the groups need not meet in expensive research facilities.

Myth: Focus group studies can be completed quickly.

Fact: Although a study can be conducted in as little as a week, even relatively simple studies typically take months to complete. While the time involved in the actual focus group interview is relatively short, considerable additional time is required in planning, recruiting, analyzing and reporting results. Clearly, focus group interviewing is a time-consuming process.

Myth: If you've been in a focus group or have facilitated other small group processes you're ready to conduct focus groups – or only highly paid consultants can do so because focus groups are so psychologically complex that only expert moderators can facilitate substantive discussion and interpret it.

Fact: Neither is true. A moderator must possess certain skills but these skills can be learned fairly quickly and honed with practice. However, some people seem to acquire these skills more easily than do others. Above all, moderators should be good listeners. People who like to talk have to restrain themselves as moderators. Moderators should be able to put participants at ease and make them willing to share their ideas. Moderators need to be assertive, yet must be able to sit back and listen. Moderators must also be diplomatic. People who possess these social skills and receive specific training in conducting focus groups make excellent moderators.

4.2 The process of conducting focus groups

To observers, conducting focus groups appears to be easy. It may seem that little effort has been expended. What is seen is the friendly and relaxed small group discussion. Not apparent are critical decisions, judgments, and planning that underlie the focus group study. It looks easy because the researcher has worked to make the situation as comfortable as possible for the participants, and the moderator is skillful in guiding the discussion. Successful focus group studies include careful designs, well thought-out questions, careful recruiting, skillful moderating, and appropriate analysis. Inattention to any one of these five factors can jeopardize the study. These are the key ingredients to high-quality focus groups.

4.2.1 Designing a study

Patton (1982:144) suggests answering six questions to outline an evaluation: What? Why? When? How? Who? and Where? These questions are also appropriate in designing a focus group study. The study design is complete when these questions have been carefully examined and answered.

1. *What* do you want to find out? On the surface this seems like a simple question, but it often requires hours of discussion over several weeks for the research team and clients to arrive at an answer. Typically this discussion period has two phases: The first phase is exploratory and seeks to incorporate different interests and concerns; the second restricts the area of inquiry, typically after decision-makers are confronted with limited resources. After discussing what it would be nice to know, the quest narrows to what you need to know.

2. *Why* do you want to know? This is the purpose of the study. It may be to describe a current situation, to generate new ideas, to make a decision, to improve a product, to garner new resources, or to add to the current body of knowledge on the subject.

3. *When* do you need the information? A deadline will help in developing a timeline for the study. Consider how long it will take to design the study, develop the questioning route, take care of logistics, recruit participants, conduct the groups, analyze the data, and prepare a report. Six weeks is about the minimum required for a study using focus groups. Occasionally it is possible to conduct a focus group study in less time, but this will typically require one or more of the following: veteran researchers, pre-identified groups, a restricted questioning route, and an abridged analysis process.

4. *How* can you get the information you need? Focus group interviewing may not be an appropriate method to obtain the kind of information you need. If you want to be able to say that 40% of a population prefers this product, don't use focus groups. Focus groups are not meant to provide statistical data. Be wary of focus group reports displaying statistics. Focus groups are used to provide insights about preferences, suggest barriers to or incentives for certain behaviors, generate or give reactions to ideas, or provide arguments for or against an idea – all from the participants' point of view.

5. *Who* is the information for and from whom should you collect it? When thinking about whom the information is for, consider both primary and secondary audiences. Primary audiences typically have one of two things in common. They have financed the study or they must make a specific decision that requires insight from the focus group study. You may have multiple secondary users: organizational decision-makers, product or program managers, elected officials, consumers, other researchers, funders. Think about how the information collected needs to be different for your various audiences. Concentrate on the needs of your primary audience.

When deciding from whom to collect the information, think about who can give you the information you need. Consider the demographic characteristics participants should have: age, income, gender, ethnic background, type of household, location of residence. Should they be users or non-users of the product? Should participants be experts on the topic? Should wholesalers, retailers, or consumers be participants? Think about how selective to be about participants' characteristics. Will you adapt your product or program for people with those specific characteristics? If not, don't segment on those characteristics. A rule of thumb is to conduct at least three focus groups with each type of person you are interested in. Therefore, if you are interested in both users and non-users and how they vary in three areas of the country, you should conduct 18 groups (3 users + 3 non-users × 3 locations).

6. *Where* do you gather the information? Think of both the geographic locations and the specific sites. Where are the people who can provide the information you need: rural or urban areas, certain parts of a country, or certain countries? There are focus group facilities located in most major cities that can be rented. These facilities often come equipped with both audio and video recording as well as one-way mirrors. But you don't need to be high tech to conduct successful groups. Community rooms, hotels, and restaurants also work well.

Practical constraints of time and money will influence your answers to each of these six questions.

4.2.2   Developing the questioning route

In designing the study you thought about what you wanted to know and why you wanted to know it. Now comes the development of the questioning route – the list of questions the moderator asks during the course of the focus group. We recommend a strategy that incorporates feedback loops between the relevant partners. Begin by inviting a few representatives of the primary information users to a special discussion. Have them brainstorm about the kinds of questions they would like answered. Have them talk about their information needs until you get a sense of what is most important. Then head back to your office and produce a draft of the questioning route. Meet again to discuss and revise the route as needed.

When you are in your office struggling with the questions, consider these hints:

1. Use open-ended questions. Instead of 'Do you like the texture?' which implies a yes or no answer, ask 'What did you think of the texture?'
2. Use 'think back' questions. Having people think about a specific ex- perience usually provides more detailed and accurate information than

you would obtain otherwise. Having them 'think back to the last time you bought ice cream' or 'think back to the last time you ate out' reminds them of a situation they can easily discuss.

3. The number of questions asked depends on several factors, including the complexity of the topic, the familiarity of participants with the topic, and the amount of time available for the interviews. Try developing 7 to 10 questions that cover the type of information required by decision-makers.

4. Start with general questions and work to the specific. The first question is unique in several respects. It is answered in order by everyone around the table to get the group better acquainted. It is non-threatening (it does not differentiate individuals by class, education, income, etc.) and can be answered in less than 30 seconds. The middle questions are transitional, leading logically to the discussion of the bottom-line questions that concern the information crucial to the study. The final question is typically general, giving participants the chance to raise any topic. 'Is there anything that we should have talked about that we didn't?' or 'What advice would you give to Sandy Jones, the person responsible for this project?' are examples of final questions.

5. The moderator may use probing questions at any time during the interview to get more description or detail. These include 'Tell me more,' 'Can you give me an example,' and 'I'm not sure I understand'. Probes are very helpful but don't overuse them.

6. Visual aids, videos, handouts, or samples may be helpful in the discussion. If you want reactions to products or programs, provide prototypes to which participants can react. If you want people to think about a number of options, use handouts or posters to illustrate them.

A good questioning route will elicit what seems like a natural conversation. One hallmark of a properly sequenced questioning route is when participants raise topics or answer questions before they are introduced by the moderator. However, when participants introduce topics or questions out of sequence, the moderator must make a tactical decision. Sometimes it is best to go with the new topic even though it is out of sequence. At other times the moderator may ask the participant to hold the idea for later discussion. The moderator can then refer back to previous comments from participants and ask for amplification.

4.2.2.1 An example of a questioning route. Fleischmann's, the University of Minnesota School of Public Health, Health Advancement Services, Inc., and the City of Phoenix worked together to develop a cholesterol management program for City of Phoenix employees. Focus groups were used in the planning stages to determine barriers and incentives to changing food habits. A screen was used to identify employees who were moderately motivated to make changes in their health related behaviors. These people were then invited to participate in the focus groups.

Questioning route for Fleischmann's Cholesterol Management Program:

1. Think back to the last time you wanted to make a change relating to health. It may be a change in what you eat, your weight, smoking or exercise habits. What kind of barriers or roadblocks did you run into?
2. If you made a change, what would have helped you the most in making that change?
3. Suppose you have been told by your doctor that he/she is concerned about your cholesterol level and wants you to make changes. What would you want to know or what kind of information would you like to get about that?
4. Some of you mentioned foods and diet. Let's focus on that. What kind of information would you want to get about foods and your diet to help you lower your cholesterol? Brief discussion, then use handout of nutrition topics for further discussion.
5. There are a lot of different ways you could get the types of information you have just been talking about. Suppose your employer, the City of Phoenix, sets up opportunities where you could get the information on diet and cholesterol. How would you like to get the information? (After a brief discussion use the handout of preferred delivery methods for further discussion.)
6. What would make the learning activity most successful for you? (Brief discussion, then use handout of success factors.) Which is most important to you?
7. As part of this program we would like to somehow involve not only city employees, but the people around them. How do you suggest we try to involve the people around you? How do we get information about cholesterol to them?
8. Suppose that a workshop on cholesterol was held on your own time. What would get you to attend?
   Probes: (a) Personal invitation from a co-worker
   (b) Competitive approach with teams of co-workers
   (c) Encouragement from your doctor, boss, friends or family
   Of all those mentioned, what do you feel is most important?
9. We realize that it's hard to stay motivated to learn about things like cholesterol, but what would entice you to come back for more? What would keep you interested?
   Probes: (a) Free coupons for healthy foods
   (b) Food tasting

    (c) Demonstrations by co-workers

    (d) Door prizes

10. We are going to be putting together programs for city employees on how to lower their cholesterol. As we begin this project, what advice do you have for us?

4.2.3 Recruiting participants

In the design of the study you determined what type of people could provide the information you need. Now you will want to locate these people, invite them to participate, and provide incentives for them to come. This is typically a time-consuming, frustrating and boring part of the focus group process. But because poorly thought-out or executed recruitment will ruin a study, pay attention to detail.

4.2.3.1 Incentives. What will get people to come? In the past many non-profits believed people would willingly donate their time whenever asked. In some cases this may be true, but most people today are extremely busy and need to see the benefit of their participation. Think about the type of people you are trying to recruit. What would get them to come? Organizations have used gifts, meals, and appeals to good will, but, all things considered, money works best. It clearly communicates to the participant that the sponsor considers the study important and respects the time of the participant. Give thought to the amount. It should not be so small that it insults potential participants. Public institutions must also be careful that it is not so large that potential participants wonder how their tax dollars are being spent. One way to gauge what the incentive should be is to ask people who fit your selection criteria. People with busy schedules and high incomes, such as physicians, will require greater incentive to participate. Light refreshments are usually provided (do not offer alcohol). Remember to offer day-care if you are recruiting parents of young children. Also remember, the greater the incentive offered the less time required to recruit participants.

4.2.3.2 Identifying potential participants. Develop a list of the characteristics required of participants. This list becomes the screen that potential participants must pass through before they are invited to participate. Screening questions may be asked either on the telephone or in person. Examples of screening questions are 'How many children do you have under the age of five living in your household?' or 'In the past week how many times have you eaten fish?' The most frustrating part of recruitment is finding people who fit the screening criteria you have outlined. The more specific the criteria, the greater the difficulty in locating participants. For example, it will be easier to find working mothers than it will be to find working mothers with incomes over $50 000 and children under five years of age.

Now you must obtain a list of names, addresses, and phone numbers of people who may meet your criteria. Think of the organizations or businesses that may have member or client lists of people similar to the type you are interested in finding. Some organizations are reluctant to share their mailing lists. It may be easier to obtain a list if a person in your organization knows someone in the other organization, personally asks for the list, explains the study, and mentions that people will receive a stipend for participation. There are also companies that sell mailing lists. Look for lists that are kept current. It is embarrassing to ask for people who have been dead for three years.

Some companies that continually test products or ideas have made arrangements with churches, hospitals, or parent associations for a supply of focus group participants. Each time a member participates in a focus group the company gives the organization a sum of money. This seems to be a mutually beneficial arrangement.

Sometimes it is difficult to find people who fit the criteria. In such cases you may want to try snowball sampling. Conduct one group and ask participants for the names of people they know who fit the criteria. Because people often associate with people like themselves, participants frequently know several people who pass the screen. Participants are usually quite willing to recommend acquaintances if they feel the incentive for participating is adequate. Get names, addresses, and phone numbers, and ask participants if you can say they gave you the name. Be sure to tell participants that you are interested in fresh insights and would prefer that they do not discuss their experience with potential participants.

Some organizations conduct on-site recruitment of participants. You may recruit participants at malls, grocery stores, schools, or parks. Occasionally newspaper ads are used. Be aware of the motives of people answering those ads (do they like to talk, want the money, have time to kill?). Random telephone screening is also a possibility, however this is extremely time consuming and should be avoided if possible.

4.2.3.3 The recruiting procedure. Begin recruiting participants about two weeks before your first focus group. Start earlier if the people you are trying to recruit are professionals with busy schedules. It is best to have the dates, times, and locations of all groups set so participants can select the group most convenient for them. Phone potential participants, ask the screening questions, and if they pass the screen invite them to a small group discussion. Typically we invite participants for a two-hour discussion and try to finish in 90 minutes. Tell potential participants what incentive is being provided. Confirm their addresses and the spelling of their names. Typically participants are given only a general idea of what the discussion is about: we will be talking about cholesterol reduction, or we will be discussing foods for children. This keeps participants from 'researching' the topic. Do not just put up a notice or send out blanket invitations to attend. These may seem like easy ways to recruit but in the end they just create headaches. You have no idea how many, if any, people will attend.

Invite seven to nine participants for each group. More than nine people are unmanageable, and it is frustrating for the participants who want to talk. Although we have conducted mini-focus groups, having fewer than four participants usually limits the breadth of opinions and interaction. If participants are experts on the subject or have a great deal of experience in the area, invite a smaller number (six or seven). These people will have much to say on the subject, hence fewer are needed for a good discussion. Over-recruit if incentives for participation are low. Invite one or two people more than you want. Don't invite more than 10 people because they may all show up.

Send a personalized letter the same day the person has accepted your invitation, confirming the date, time, and place of the focus group, and including a contact to call if they are unable to attend.

Phone each participant the day before or the day of the group to remind them of the discussion. This recruiting procedure may seem like overkill but it increases attendance rates considerably. People have been contacted about the discussion three times. It signifies that this is an important discussion.

4.2.4   Moderating

Myril Axelrod, a market researcher who has written much about the use of qualitative methods in marketing, describes the selection of a moderator:

> In selecting the moderator the primary requirement is how well that person is going to be able to listen – is he or she someone who is really interested in people, who wants to hear what someone else has to say, who can readily establish rapport and gain the confidence of people, who can make them feel relaxed and anxious to talk? This can only happen if the moderator is genuinely that kind of a person. (Axelrod, 1975, p. 6)

The primary prerequisite for moderators is that they be good listeners.

Select moderators who appear to be similar to the participants. The more the moderator appears to be like the participants, the easier it is to build trust and make people comfortable. The gender, age, race, or ethnic background of the moderator may influence what participants say. For example, if you are studying the food preferences of Hispanics consider a Hispanic moderator.

The moderator should have some background in the topic to be discussed. Although some naïveté on the moderator's part may elicit more explanation from participants, being too naïve is offensive to people who have taken their time to meet to discuss the topic. This is particularly true for people with a high level of expertise. However, we have also found that it is best to have a moderator not closely associated with the project. People with a vested interest have difficulty hearing anything but what they already believe. They may dismiss or undervalue perceptions, and sometimes they feel the need to defend, explain, or educate – all tactics that jeopardize the focus group. Moderators without personal interest in the topic are better able to listen.

4.2.4.1 The assistant moderator. This individual sits in on the focus group and helps with specific tasks. The assistant moderator does not participate in the discussion, but typically sits back from the table, observes, takes notes, and runs the tape recorder. The assistant helps set up for the group, and greets and orients latecomers. Near the end of the session the moderator asks the assistant moderator if he or she has any questions for the group. At this time the assistant moderator may begin to participate in the discussion. Involving the assistant before this often confuses the group. The most valuable role of the assistant moderator is in the debriefing session after the discussion when the moderator and assistant moderator share observations. Assistant moderators are not often used in commercially conducted focus groups because of the added cost, but we believe the assistant moderator can add greatly to the quality of a focus group.

If program managers or decision-makers act as assistant moderators or sit in on focus groups, make sure they observe a series of focus groups and avoid the temptation to rush to a decision based on one group interaction. If they view only one group, they don't have an opportunity to observe trends. They may fasten onto a remark or idea that is not representative of the group discussions.

4.2.4.2 Preparation just before the session. The moderator must prepare for each focus group. Moderating requires energy. Be well rested and mentally prepared for the session. Don't plan to do more than two groups per day. Even veteran moderators have difficulty concentrating on the flow of the discussion when doing more than two groups in a day. Momentary lapses of memory by the moderator can be hazardous and embarrassing (was this just discussed in this group or was it in the group before this?) and can be avoided by limiting yourself to two groups per day.

Know the questioning route well. Questions should flow smoothly from the moderator. Don't read the questions and don't spend much time looking at the questioning route. The process should seem like a discussion, not an interview.

Check materials and equipment. Make sure you have a list of names and phone numbers of participants, name cards for participants, twice the number of audio-tapes you actually need, tape recorder, microphone, extension cords, extra batteries, and any other materials you are using in the group. Check all equipment to make sure it works.

Arrive at the meeting site at least 30 minutes before the session is to begin. Set up the seating arrangements. The moderator sits at the head of the table. The assistant moderator sits opposite the moderator but away from the table. Participants sit in a circle around the table. You may also need to check on refreshments or request changes in the room temperature, lighting, noise level, etc. You also need time to set up your equipment. All this should be accomplished before participants start to arrive.

4.2.4.3 During the session.   As participants arrive, offer them refreshments and engage them in small talk. Talk about weather, sports, or traffic but avoid talking about the topic you will be discussing in the group. You want to capture perceptions in the group discussion, and if participants discuss the topic before the group begins, they may be reluctant to share similar information later in the group. If you want background information, have participants complete these forms before the group discussion begins.

When most participants have arrived or when a reasonable amount of time has elapsed after the announced starting time, begin the group. The assistant moderator can meet latecomers at the door, give them a short briefing, and bring them into the group. Give the introduction without looking at your notes. Your job is to create a comfortable environment. Be sincere. Be friendly. Make the participants feel that they can trust you and that you want to hear what they have to say. Make eye contact (if culturally appropriate). Smile. Put participants at ease.

The moderator asks each participant to answer the first question, usually by going around the table. The earlier a person talks in a group, the greater the likelihood that they will talk again in the session. For the rest of the session participants are encouraged to speak whenever they want. The moderator will become less central to the discussion as participants begin to build on each other's comments.

Beginning moderators wonder if they should deviate from the questioning route within a focus group. At times participants will raise a subject that was not anticipated. The moderator may wonder whether to follow this up or to let it go. These serendipitous questions can often lead you astray and consume precious time. It is best to note the topic and come back to it at the end of the session if time permits.

Two problems torment moderators: people who don't talk and people who talk too much. Quiet people are easier to deal with diplomatically. Here are a few techniques that may encourage quiet participants to share their insights:

1. Eye contact. Typically eye contact will encourage people to talk. It is a way of giving people permission to speak. If you know or suspect a participant is quiet or shy, seat them opposite you at the table. When you look up, you will be looking at them. Some people dislike eye contact and will look away, but most people will assume they should say something.
2. The five-second pause. People feel uncomfortable with silence in a group and will try to fill the void. The moderator can take advantage of this by purposefully not talking. The pause gives participants an unambiguous message that their insights are wanted. Incidentally, the pause is most effective when combined with eye contact.
3. Call on participants. If the first two techniques don't work, you may want to tactfully call on participants. Saying 'Janet, I feel we haven't given you a chance to talk . . .' or 'Bob, I'd like to hear your opinion' lets quiet participants know you want to hear from them.

Other participants take great pleasure in talking. They talk often and at length. They feel they are more knowledgeable about the discussion topic than others in the group and therefore, as an expert feel compelled to share their insights. Others who talk often will ramble. They seem to think and talk simultaneously and they aren't sure where the idea will end until they have heard themselves say it out loud. Whatever the reason, these individuals present special concerns in a focus group. If you know or suspect that a participant is a talker, try to seat them to the immediate left or right of the moderator. This makes eye contact with the moderator difficult. If someone is rambling, stop looking at them, look down at your papers but *don't* take notes. The talker may view notetaking as encouragement. If the talker persists in dominating the conversation interrupt and say 'Thank you, I'd like to hear from some of the other people now' or 'Let's give some of the other people a chance to talk'. Be diplomatic. You don't want to embarrass or offend, but you do need to be assertive.

There are several ways to end a focus group session. One for the moderator to thank people, take care of any paperwork (like incentives), and send them on their way. Another method is to have the moderator or assistant moderator summarize the discussion and have the participants comment on the adequacy of the summary. It is hard to summarize a two-hour discussion in a few minutes, but it is valuable in later analysis and also provides a check on what you heard.

4.2.4.4 After the session. Check the tape recorder immediately after the session to make sure the discussion was recorded. If for some reason it wasn't, immediately write down everything you can remember about the discussion. If you had an assistant moderator taking notes, this should be fairly easy. If the recorder is working, turn it back on and talk about what you heard, observations, hunches, interpretations. Take at least twenty minutes to do this. Draw a diagram of the table and list the names of the participants to use for future reference. As soon as possible, write a summary of the group session. Don't assume you'll remember what happened when you are ready to formally analyze the data. Memory fades especially after participating in several focus groups with similar audiences. Make a copy of the audio-tape. If you are having transcripts typed, send the tape to the transcriptionist. Transcripts make analysis easier, but unfortunately they are time-consuming and costly. It takes a good typist about 12 to 16 hours to transcribe one focus group, so give your typist the tapes as you conduct the groups. If you wait until all the groups are completed, the task will seem overwhelming. Make sure support staff are aware of this project well in advance.

4.2.5 Analysis

Analysis can be one of the most vexing and suspect aspects of conducting focus groups, particularly for people who don't have a background in qualitative research. Six hours of audio-tape or 100 pages of transcripts can be overwhelming if you don't have a good idea of how to proceed. In this section, we will summarize some characteristics of qualitative analysis and outline a process for analysing focus group data.

4.2.5.1 Qualitative analysis. Qualitative analysis, whether you are reviewing documents, conducting individual or group interviews, or analyzing written responses, has these characteristics:

1. It takes time. The amount of time required depends on your experience with analysis, the analysis procedure used, and the type of information you are interested in. As the research question becomes more complicated or the decisions to be made more important or costly, the analysis becomes more rigorous and, therefore, time-consuming.
2. It is ongoing. It begins when you start collecting data – with your first interview or document. It continues throughout the data collection as you look at each new bit of information to determine how it is similar to or different from earlier data.
3. Teamwork helps. It helps to have a colleague with whom you can discuss your research. It is even better to have a team of researchers on a project. Throughout the analysis process you will be weighing findings, speculating about interpretations, and formulating potential recommendations. Discussion with another person often helps clarify

and expand thinking, and it helps to minimize the incorporation of personal biases.

4. It must be systematic and verifiable. Analysis must follow a prescribed, sequential process. There are different processes that may be used, but, whichever process you choose it must be systematic. Follow a predetermined process and be able to describe that process to another person. The analysis must also be verifiable. This means that another person should be able to arrive at the same findings based on the data. Of course, another person may see additional findings, or have alternative interpretations or different recommendations, but the findings should be verifiable by another researcher.

4.2.5.2 One analysis process. There are many ways to analyze focus group interviews. We have found that after people have completed several studies they tend to develop their own method. We suggest you begin by using the method recommended here. It adds structure to what can otherwise be a vague process. Later we will briefly describe some alternatives. After you have experience analyzing focus groups you may want to try other methods.

Should the moderator and the analyst be the same person? This isn't necessary, but analysis is easier if the analyst has sat in on the groups. By observing or participating you directly experience the situation with all your senses. You get a feeling for what is happening that is difficult to obtain from listening to audio-tapes or reading transcripts.

Here is our recommended process for analysis:

Step 1. Assemble all your research materials: questioning route, audio-tapes, assistant moderator notes, summaries of each group (completed immediately after each focus group), hard and disk copies of transcripts. For certain studies you may have other materials to be used in analysis such as background data on participants, written comments collected from the group participants, or flip charts used within groups.

Step 2. Read all the summaries at one sitting. This will provide a quick overview and refresh your memory about the setting, the participants, the tone of the discussions, and general reactions to the discussions.

Step 3. Listen to the audio-tapes. If you did not witness the focus groups, listen to the tapes. This can be time-consuming, but it gives you a better idea of what was happening than does just reading the transcripts. Note the level of energy, who is talking, where the discussion is stilted. If you moderated or assisted in the groups, you can skip this step.

Step 4. Read each transcript. Get a general sense of what is being said.

Jot down in the margin ideas or potential patterns. Use high-lighters to mark important words, phrases, or quotes.

Step 5. Examine responses to one question at a time. This makes analysis manageable. For example, look at all the responses to question one across each of the groups. Look for patterns or themes across the groups. Look for similarities and differences among responses. Here are some suggestions to guide your analysis:

(a) Examine the words. Certain words convey a vivid picture of what respondents are saying. In one study we conducted, a participant said he was tired of having new programs 'dumped' on him by administration. The word 'dumped' connotes a strong feeling about the way programming decisions were being made.

(b) Consider the context. Look at what is being said before or after important statements. Does a particular response seem to spur another?

(c) Look for the presence or absence of internal consistency. Consistent responses are easier to analyze than inconsistent responses. If people are changing their opinions within the group, try to find out what is prompting the change. Does a particular argument or statement seem to persuade people to another point of view?

(d) Consider the specificity of responses. Place more emphasis on responses that are specific. People who have had experience with your topic will be able to give you more specific, and typically more helpful, answers.

(e) Consider the frequency/extensiveness of comments. Place more emphasis on comments said more frequently and said across groups. Frequency is not the only factor to consider, however. If only one person says they would buy this product, but they say it repeatedly, place less emphasis on this than if a number of people within each group make the statement once. It is important to determine the extensiveness of comments. Sometimes assistant moderator notes of apparent agreement (such as head nodding) can add indicators of extensiveness of agreement. At times a comment is made by only one person, but it is an important insight which has a bearing on the study. The analyst may want to cite the comment but indicate that the view was expressed by only one participant.

(f) Consider the intensity of the comments. Gauge this through participation in the groups or listening to the audio-tapes. Indicators of intense feelings may appear as changes in pitch, tone, or speed of speech, or physical gestures such as pounding on the table.

Step 6. Look for big ideas. Look across the answers to individual questions and across groups for themes that continually appear. Specific comments may, when considered together, combine into a broader idea. There may be an umbrella theme that encompasses smaller themes. In focus groups conducted to develop a cholesterol reduction education program for City of Phoenix employees, one participant said 'I don't make anything that takes longer to cook than it does to eat'. Another talked about not wanting to clean up the mess created in cooking most meals. Someone else talked about not wanting to search for the exotic ingredients needed for healthy recipes. Throughout the interviews people said things related to time and convenience. Program designers knew that to be successful the new program must incorporate and emphasize convenience. It became a critical dimension of the program.

4.2.5.3 Other analysis procedures. The analysis procedure chosen is often related to the importance of the decisions to be made based on the findings and the amount of resources available for the study. Consider the cost or reversibility of the decision to be made, the complexity of the problem being studied, and the amount of time or money available for the study when choosing an analysis procedure. The same procedure is not appropriate for all occasions. We estimate that the analysis procedure just outlined requires one week for each series of three focus groups. There is a more elaborate analysis procedure that involves coding transcripts on a computer. Even more elaborate is coding by several analysts, checking for inter-rater reliability, and recoding. These methods provide more detail than is required for most studies. Here are some other options:

Analysis from abbreviated transcripts. If you don't have funds for transcripts but you have time, try typing abbreviated transcripts. Transcribe each tape immediately after the session, but transcribe only those parts that you think will be pertinent to the analysis. Eliminate the introduction, rambling comments, and anything off-topic. Then conduct the analysis as described above.

Analysis from audio-tapes. For some studies transcripts are a luxury that cannot be afforded. In this method the analyst forgoes transcripts and conducts the analysis by listening to the audio-tapes and examining the group summaries. Typically the analyst will transcribe quotes and make notes on a word processor while listening to the tapes. This method eliminates clerical transcription time and associated costs but may increase analysis time. Comparison of responses between groups becomes more difficult because the analyst cannot easily move from one group to another.

Analysis from summaries. The moderator or assistant moderator uses assistant moderator notes and group summaries for analysis. The analyst typically reviews these materials and notes prominent general themes of the discussions. It is best if the moderator and assistant moderator work together on this type of analysis to minimize the effects of a selective memory.

Computer-assisted analysis. If you are looking for a software program to do analysis for you, forget it. However, programs are available for both IBM compatible and Macintosh personal computers to help you manage your data, making analysis more systematic. Their main function is to assist in the coding and sorting of data. The cut and tape of the past is now done by computer. The analyst codes sections of text which the computer can group or retrieve. For example, the analyst can code interviews by age and then pull all responses of teenagers and compare them to responses of people in their forties. Or all answers to question three can be examined at the same time. Or all passages dealing with food texture can be retrieved. These programs are particularly helpful if you have large amounts of text, which are difficult to conceptually compartmentalize all at once.

For more information about software programs for qualitative analysis and their relative advantages and disadvantages, consult *Qualitative Research: Analysis Types and Software Tools* by Renata Tesch (1990).

4.2.6 Validity and reliability – Can we really trust this stuff?

Researchers thinking of using focus group interviewing for the first time often have questions about validity, reliability, and generalizability. Most researchers have been trained in positivitistic research techniques for which random sampling, researcher distance, and standard statistical procedures for analysis are the norm. Focus groups have none of these, which makes researchers nervous. However, 'regardless of the type of research, validity and reliability are concerns that can be approached through careful attention to the study's conceptualization and the way in which the data were collected, analyzed, and interpreted' (Merriam, 1988, p. 165).

Qualitative researchers incorporate a variety of techniques to ensure internal validity or credibility. These techniques include:

1. Long-term observation: Spending sufficient time in data collection to determine what information is most relevant to the situation and to overcome misinformation or distortion.
2. Peer debriefing: Discussing findings and interpretations with a colleague.
3. Member checks: Asking participants to review the findings and interpretations for plausibility.
4. Monitoring researcher biases: Writing down biases, assumptions, and hypotheses at the beginning of the study and archiving them. Later the researcher checks findings and interpretations against these initial hunches. A researcher may find that he or she is resisting new ways of thinking about the situation because of preconceptions.
5. Triangulation: Using multiple methods, multiple sources of data, or multiple researchers to get a more complete picture of what is being studied.

Some beginning analysts are concerned that another researcher conducting the same study will come up with different results. Actually, it is to be expected that someone else would arrive at different results since data collection and analysis are dependent upon the researcher's skill in obtaining, synthesizing, and reducing data and their background. Researchers filter information based on their knowledge of and experience with the topic. Different researchers may see different parts of the problem. They will place varying importance on the data. Qualitative researchers are not overly concerned about this since it is seen as multiple interpretations of the situation. Differing results are often actually complementary. They provide differing views of the same situation. Instead of being concerned that someone else will come up with differing results, the researcher should be concerned that the data collected lead to the results described.

As for generalizability, a number of researchers believe it is an inappropriate concept in qualitative research since criteria to determine this deal with large sample size and random samples which are typically not used in qualitative research. An alternative view that has received acceptance is that of reader or user generalizability. The researcher does not make generalizations – the reader does. This requires that the researcher provide enough description of the participants, the setting, and the environment of the discussion that the reader can determine if the results apply to a given situation. Therefore, when presenting findings it is important for the researcher to provide the types of information decision-makers will need to make transferability decisions. Egon Guba and Yvonna Lincoln offer an excellent discussion of this topic in Chapter 8 of *Fourth Generation Evaluation* (1989).

By building in techniques to ensure credibility and providing enough detail for users to determine if results are applicable to another situation you can address some of the concerns about validity and generalizability.

` 4.3 Summary

Focus groups offer a unique research methodology that can complement other types of food preference research. The technique enables researchers to get close to the customer and obtain information about preferences and behaviors. As is true of all social science research methodology, focus groups have advantages and limitations, and they have been subject to abuse. Poor planning, inadequate resources, lack of attention to the questions and the way they are presented are vexing concerns common to all research. This chapter highlighted five of the critical factors for successful focus groups: careful design of the study, well thought-out questions, careful recruiting, skillful moderating, and appropriate analysis.

References

Axelrod, M.D. (1975) 10 Essentials for good qualitative research. *Marketing News*, March 14, pp. 5–8.
Guba, E.G. and Lincoln, Y.S. (1989) *Fourth Generation Evaluation*, Sage, Newbury Park, CA.
Krueger, R.A. (1994) *Focus Groups: A Practical Guide for Applied Research*, 2nd Edn, Sage, Newbury Park, CA.
Merriam, S. (1988) *Case Study Research in Education*, Jossey-Bass, San Francisco, CA.
Patton, M.Q. (1982) *Practical Evaluation*, Sage, Beverly Hills, CA.
Tesch, R. (1990) *Qualitative Research: Analysis Types and Software Tools*, The Falmer Press, Bristol, PA.

5 Product optimization: approaches and applications

H.R. MOSKOWITZ

5.1 Background and applications

5.1.1 What is product optimization?

In its most general sense, the phrase 'product optimization' stands for the disciplined approach to product development, whereby the investigator systematically varies formula and processing conditions. In recent years product optimization has become an increasingly valuable tool for development, because it cuts time, cost and risk in the development process (Baxter, 1989; Box *et al.*, 1978; Gordon, 1965). This paper presents a history of the approach from the viewpoint of applications, followed by an illustrative case history.

5.1.2 Historical background

Thirty years ago, in the mid 1960s, manufacturers were content to develop products by methods they considered 'tried and true'. These methods were often 'rifle shots', wherein the developer submitted his or her best guess to the product evaluator. The concept of systematic product variation as an aid to development had not yet captured the attention of product developers, although from time to time one or another article on experimentally designed products would appear in the scientific literature (e.g. Gordon, 1965). Computation facilities were primitive by today's standards, and comfort with statistical procedures was minimal. Statistics, in the 1960s, primarily involved inference to determine whether two samples came from the same sampling distribution. Modeling consumer reactions and relating these reactions to systematically varied formulations was rare, although research interest did focus on the quantitative relation between subjective reactions and physical measures of product characteristics. Psychophysicists, however, were the principal researchers who focused on the quantitative relation between magnitude of stimulus level and intensity of subjective response (Stevens, 1975).

Looking back at the scientific literature from the 1930s to the 1970s we may well be amazed that most basic and applied researchers consistently ignored the power and benefits of systematic product design and optimization. As is often the case, however, the zeitgeist of the time did not focus on improving ways to develop products. The late 1940s and 1950s had witnessed the explosive growth of product development to feed nations starved for new products, first because of a deep economic recession, and second because of a war which had diverted material and manpower to secure military objectives. By the late 1960s, however, the era of opportunity began to wane. Fewer and fewer truly new products were being developed as a proportion of actual products introduced to the marketplace.

It was with the widespread use of computers, first as mainframes, followed by personal computers and workstations, that experimental design and optimization came into its own. The economic climate of the 1970s and 1980s appeared to promise endless growth, so researchers did not perceive the clear need to improve product development procedures. Despite the deceptive impression of *status quo* on all fronts, however, the availability of increasingly cheaper computing made experimental design ever more attractive. Laboratory researchers in both government and industry took tentative steps, applying experimental design to small-scale practical problems. Typically, these initial studies involved one to three physical factors, systematically varied over a modest range. Experimental design quickly proved itself valuable as a teaching aid. It showed what effects occurred by changing variables, and what interactions existed among two or three variables.

Each scientific discipline comes bedecked (and encumbered) with its own array of symbols, language, and metaphors. Scientists and statisticians involved in experimental design and modeling are no exception. The early work in experimental design pioneered by statisticians was relatively free from fixed ideas about what was 'correct' or 'incorrect' in systematically varied test designs and analysis. The primary focus was to encourage people to use a design rather than rely on unconnected 'rifle shots'. With increasing sophistication, however, focus soon shifted to the proper design and appropriate analysis. Specifically, what was the appropriate way to analyze the data? Depending upon who created the test design and developed the ensuing model and optimization, one might wind up with one of two analytic approaches: dose response functions versus iso-response contours.

5.1.2.1 'Dose-response' functions. The first approach favored by scientific researchers examined the relation between a single variable and a single response. The researcher would first develop a model relating the full set of independent variables (and their interactions) to a single dependent (e.g. overall liking). Typically, the relation took the form of an equation (sometimes linear, more often parabolic or second order). The equation embodied several variables simultaneously (viz. those that the investigator had systematically varied). The researcher would then hold all variables constant but one, systematically vary that single variable over a range, and estimate the rating that a panelist would assign to the product. Figure 5.1 shows a 'sensitivity analysis' or 'dose-response' curve.

Curves such as that shown in Figure 5.1 were promoted by researchers trained in psychophysics, a branch of experimental psychology. Figure 5.1 shows the relation between one physical variable and one response (taking into account the effects of the other variables, which for the analysis are held constant).

5.1.2.2 Iso-response contours. The second approach, favored by statisticians, consists of creating a set of charts, or iso-response contours, like those shown in Figure 5.2. Figure 5.2 uses the same equation relating the physical variables to the response. It does not look at one independent variable at a time (versus the response), however. Rather, Figure 5.2 looks at pairs of variables at one time (holding the remaining variables constant). It develops contours – combinations of the two variables which simultaneously generate a fixed response.

Figure 5.2 embodies the viewpoint of statisticians, interested in general patterns in the data, rather than the viewpoint of experimental psychologists interested in the effect of each variable.

An optimum for the experimental psychologists comprises that set of ingredient levels maximizing the response. An optimum for the statistician is the contour that is 'highest' in terms of the response. In a sense, the experimental psychologist is interested in the *one point* that maximizes a response, whereas the statistician is interested in *the region* or combination of variables which generate equally high (and near optimum) responses.

These are two historical approaches. Which is right? How have things changed up to the present day?

5.2 Steps in a designed experiment and product optimization study

Currently, investigators follow a standard sequence set of steps to optimize a product. Although variations in the sequence occur, virtually all studies follow a simple format design which goes from formula development to testing to analysis:

Step 1. Selection of relevant variables and layout of levels by experimental design.

Step 2. Questionnaire development for testing among respondents (and/or parallel instrumental measures).

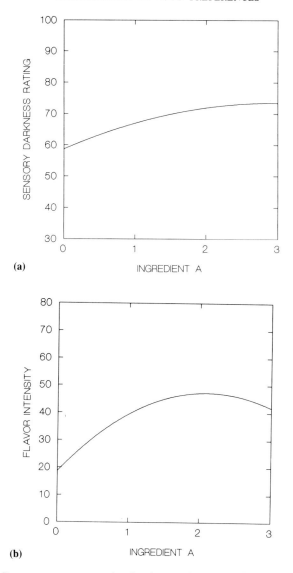

Figure 5.1 Dose response curves showing how an ingredient (the independent variable) drives a sensory response. (a) Darkness versus ingredient A. (b) Flavor intensity versus ingredient A.

Step 3. Test implementation.
Step 4. Data analysis and database development.
Step 5. Creation of models from the empirical data.
Step 6. Use of the models for predictions of product fitting specific goals.

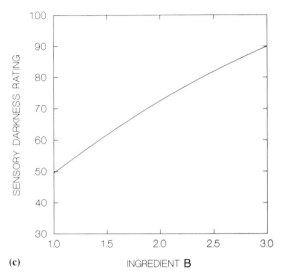

Figure 5.1(Continued) (c) Darkness versus ingredient B.

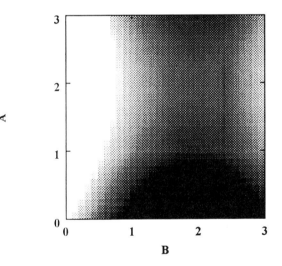

Figure 5.2 Equal intensity contour, showing how two ingredients, A and B, jointly determine 'overall liking'. Darker area signifies greater degree of liking.

5.2.1 Selection of variables and their levels – systematic versus haphazard designs

The first step identifies the relevant independent variables. Because modeling and optimization depend heavily upon equations relating independent variables to consumer responses, it is best when the investigator selects the independent variables ahead of time and arranges their levels according to an experimental design. An experimental design (in contrast to haphazard design) arrays the independent variables so that the variables are 'statistically independent' of each other. Statistical independence enables the investigator to measure what independent variable separately contributes to the consumer response, and add together the weighted effects of these independent variables. In contrast, when the investigator fails to systematically array the independent variables in an experimental design, quite often the independent variables covary with each other. Changes in one variable are confounded with changes in another. A haphazard design makes it impossible (or at least extraordinarily difficult) to tease out the effect of each variable separately.

5.2.2   *Questionnaire development*

Questionnaires link consumers and researchers. The questionnaire requires the panelist to assess the product on a variety of characteristics. If the questionnaire contains the proper attributes with a sufficiently sensitive scale, then the experiment produces usable data. If, perchance, the investigator asks the incorrect questions and misses key attributes, then the results will be worthless no matter how well the study is designed statistically and executed in the field.

Questionnaires, like scales, generate a lot of heat and occasionally some light. Researchers from opposing camps hold different points of view about what is appropriate to ask a consumer and what is not. (More often than not, however, these opinions are neither supported by experience nor data. They remain opinions.)

At its most generous, the questionnaire requires the panelist to rate a product on a full array of attributes. These attributes include sensory ratings, liking ratings, image ratings and directionals.

5.2.2.1 *Sensory attributes.*   Sensory ratings are ratings of the degree to which a specific characteristic is present in the product. Examples include perceived darkness of the product (from 'light' at one end of the scale to 'dark' at the other end of the scale). The assessor makes no judgment about whether the sensory attribute is 'acceptable' or 'unacceptable', or whether the attribute is too weak, just right, or too strong, relative to what the assessor desires. Rather, the assessor acts as an instrument (albeit a subjective one), registering the amount of a characteristic perceived in the product.

Assessors can easily rate sensory attributes, especially when these are explained to them during an orientation. The scientific literature of psychophysics in experimental psychology demonstrates conclusively that assessor ratings 'track' physical changes, when the assessor is presented with systematically varied stimuli varying along a single dimension and instructed to scale the corresponding sensory attribute (e.g. sweetness of cola versus known variations in sugar level; Moskowitz *et al.*, 1980).

In product evaluation the assessor can rate a full profile of sensory characteristics, including appearance, aroma, taste and texture. For the questionnaire to be of maximum benefit the assessor should rate specific attributes of a sensory dimension as well as the overall intensity of that dimension. For instance, if the product is a beverage and the attribute is taste intensity, then the assessor should rate overall taste intensity as well as the separate intensities of sweetness, bitterness, and tartness. Assessors experience no trouble rating both the overall intensity of an attribute, and the more familiar sensory aspects (e.g. sweetness). Assessors may have trouble rating some of the finer and esoteric nuances of the sensory dimension (e.g. woodiness of fragrance), simply because they do not know what the attribute means. A short explanation of the attribute and/or reference standards can help to clarify what an attribute means.

5.2.2.2 Liking attributes. Liking attributes require the consumer to make a different judgment – whether the product is acceptable or unacceptable. Assessors have little trouble assigning ratings to liking, because for food products liking is a primary response (especially for flavor and aroma characteristics; Yoshida, 1964).

Liking questions come in many forms. As in sensory attribute ratings, the assessor may rate overall liking as well as its component attributes (e.g. liking of appearance, liking of aroma, liking of taste, liking of texture or mouthfeel). Assessors easily rate overall liking. For most foods, assessors can also easily rate liking of sensory characteristics when those characteristics are expressed in the most general form (e.g. liking of appearance). Occasionally, however, researchers are carried away with the questioning process and ask more detailed liking questions, such as 'liking of the dark color', or 'liking of sweetness'. For such highly specific questions, liking ratings are more difficult. Although assessors have no trouble rating sweetness, many panelists report having difficulty rating 'liking of sweetness'. The liking ratings simply do not mean much at such a 'micro level'. (Despite the difficulty, however, many researchers put these questions into a questionnaire, time after time, and obtain consumer compliance and numerical ratings.) Assessors interviewed at the end of a test session often report that they find it hard to separate liking ratings of a specific nature (liking of saltiness, for instance) from the general attribute liking (liking of taste).

Investigators use other scales besides amount of liking. The panelist can classify the product as 'acceptable' or 'unacceptable', 'satisfactory' or 'unsatisfactory'. These are 'all-or-none' judgments (classifications). The

Table 5.1 Verbal descriptors for hedonic scales

Scale points	Descriptors
2	Dislike, unfamiliar
3	Acceptable, dislike, not tried
3	Like a lot, dislike, do not know
4	Well liked, indifferent, disliked, seldom if ever used
5	Like very much, like moderately, neutral, dislike moderately, dislike very much
6	Very good, good, moderate, tolerate, dislike, never tried
5	Very good, good, moderate, dislike, tolerate
5	Very good, good, like moderately well, tolerate, dislike
9	Like extremely, like very much, like moderately, neither like nor dislike, dislike slightly, dislike moderately, dislike very much, dislike extremely

Source: Meiselman (1978).

assessor does not act as a measuring instrument which assigns intensity or magnitude ratings to the degree of liking. Rather the assessor assigns the product to one of two (or one of a limited number of) categories. Analysis of classification data shows the percent of panelists who classified the product into each category (rather than the average rating assigned by the assessor for the product, along the specific attribute). The assessor can rate 'purchase intent', defined as either the probability of purchasing the product, or selecting the proper category which defines the consumer's purchase intent (e.g. from 1 = definitely not purchase, to 5 = definitely would purchase).

5.2.2.3 Image attributes. In recent years marketing researchers have expanded their use of image laden attributes into product testing and product optimization. Image attributes are characteristics which transcend simple sensory impressions or liking. Image characteristics call into play sensory impressions of the product along with needs and expectations. For instance, for beverages, image characteristics might be terms such as 'natural', 'refreshing', 'caloric', 'youthful', etc. The image attributes are not necessarily sensory, because they do not invoke simple sensory reactions, although they do use sensory impressions as inputs. Image attributes are not liking attributes either, although quite often the image attributes correlate highly with liking. (The correlation is not always perfect. For instance, the attribute 'unique product' is an image attribute. Often overall liking first increases with increasing 'uniqueness', peaks and then drops down as the rated uniqueness of a product exceeds some optimal level.)

5.2.2.4 Sensory directional attributes. Many product testers instruct the assessor to rate the degree to which a product under-delivers, or over-delivers a specific characteristic. For instance, with processed meat the assessor might be instructed to rate saltiness as a directional attribute, evaluating the sample as having a 'too weak' salty taste, a 'just right' salty taste, or a 'too strong' salty taste. Directional attributes are best used when the assessor evaluates one or two disparate, unconnected products. The product developer uses the directionals to figure out what is 'wrong' with the product. In studies using systematically designed products the sensory directional ratings are usually less valuable.

5.2.2.5 _Scales._ In product optimization almost any scale suffices, as long as the scale differentiates among products on the key attributes. The goal of product modeling and optimization is to create an equation relating formula level to ratings. The scale is merely a device to show differences among products.

Researchers have used a variety of scales in optimization, studies including short scales (e.g. 5-point rating scales), long scales (e.g. 0–100), open ended ratio scales (viz. magnitude estimation), and so forth. It is not clear that any one of these scales is particularly better than another for use in the optimization procedure. However, over a number of years psychophysicists interested in the mechanisms of perception have opted to use wide scales, e.g. magnitude estimation (Stevens, 1975). These researchers aver that the wider scales enable the assessor to assign numbers more 'accurately' to the products, which properly becomes important when the research goal is a model.

Often researchers use percentage data from scales, rather than the scale values themselves. For instance, instead of computing a mean from a 5-point liking scale, the investigator can compute the percentage of consumers who rate a sample as '5', '4', '3', etc., and can even aggregate the percentages from two or three different categories. A good example of this scale analysis is the 'per cent top two box' purchase value, comprising the proportion of consumers who rate a product as 'definitely would buy' or 'probably would buy' on the 5-point scale. These statistics are 'incidence' values, or percentages of consumers who assign specific responses. (Percentage statistics can be created for any part of the scale, such as the percentage of consumers who would probably or definitely _not_ purchase the product.)

5.2.3 Test implementation

Product optimization studies always involve multiple product formulations. The studies allow evaluation of these formulations, along with 'benchmark' products (e.g. competitors, tested in the same design).

The field specifics of the study entail a number of decisions, which impact on the information that one can obtain and the conclusions that one can draw. Some of the issues are explicated below.

5.2.3.1 *Where is the test conducted?*

Optimization tests can be conducted in one of two different venues – central location (supervised), and home use test (usually unsupervised). In a central location test an interviewer prepares the product(s), under controlled conditions, gives the sample to the assessors, and obtains the assessor's ratings. The central location test is conducted under careful supervision. The product preparation can be controlled to the most demanding specifications (within the ability of the field service executing the project).

In a home use test the assessor rates the product after (preparing and) consuming the product at home. There is no supervision. The assessor may rate the product just after preparing and consuming it, or may wait until an interviewer calls and then rate the product from memory. Home use tests provide a more natural environment for product evaluation. Usually, the assessor evaluates only one product (or a few over an extended study) rather than evaluating several products (as is done in the central location test procedure).

5.2.3.2 *How many products does the assessor rate, and in what format?*

Central location tests often collect multiple product ratings from a single assessor. The supervisor at the test site orients the assessors, and presents them with the stimuli in a randomized order. In an extended test session lasting several hours, an assessor who has been oriented in evaluation can test 2–10 or more products, with little sensory fatigue. As long as the assessor knows what to do and waits at least 5+ minutes between evaluations, he or she can evaluate many different products in the session. (Parenthetically, with highly intense flavors, the assessor may have to wait longer than 5 minutes to recover.)

In home use tests, without any supervision, the assessor is limited to fewer products. In the standard home use design an assessor uses a single product for a week or more and then completes a questionnaire. Home use formats require the assessor to use one, or at most two products over an extended period. The nature of the field work in conventional home use tests makes optimization research harder and more costly when there are many samples to study. With the lack of control and with minimal number of products tested by each assessor, the experimental design for products requires many more assessors. (For instance, for 15 products, each tested by a total of 50 assessors, and with each assessor testing two products, the total number of products × assessor ratings = 750 (15 × 50), and the total number of assessors needed if each assessor evaluates two products is 375 (viz. 750/2). In contrast, in a central location test

Table 5.2 Benefits and drawbacks of testing procedures

Central location tests	
Benefits	Drawbacks
1. Supervision 2. Careful product preparation 3. Many products can be evaluated in a short period of time	1. Not a natural environment 2. Products may not be prepared in a manner that the respondent is accustomed to. 3. People can be affected by other respondents. 4. Cannot measure long-term responsiveness to a product.

Home use tests	
Benefits	Drawbacks
1. More natural environment 2. Can measure long-term responsiveness to a product	1. Unsupervised rating of the products and product preparation 2. Many products cannot be evaluated over a short period of time

where each assessor rates 10 products, the total number of 'testings' is still 750, but only 75 assessors are needed.)

An alternative method, preferable for product optimization, allows the assessor to evaluate multiple products at home during the course of a week. The assessor takes home two to five products for the week, evaluates one product per day (preparing the product and consuming the product as the assessor normally would at home). Just after preparing and consuming a specific product the assessor rates that specific sample. The study can go on for several weeks, during which time the assessor rates multiple products.

Central location tests and home use tests have different benefits and drawbacks. These are listed in Table 5.2. Overall, neither test is 'better' than the other. Both measure reactions to the products. Home use tests can measure long-term responsiveness to a single product, whereas central location tests cannot.

5.2.3.3 How many assessors should participate? The 'base size' of assessors participating in a study depends upon the nature of a study. Product optimization research often requires fewer assessors per product than do conventional experiments which measure reactions to one or two products. Optimization studies seek patterns relating physically varied stimulus characteristics to subjective responses. It is not necessary to be 'overly precise' in the measurement of one product, because the pattern of responses to the stimuli can still be clearly determined even if each stimulus is measured with a moderate amount of 'noise'.

Consider the results in Figure 5.3(a), which show how liking varies with spice level in salsa. Despite the fact that there is substantial variability

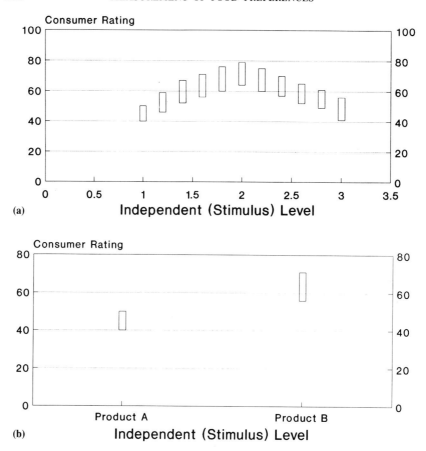

Figure 5.3 (a) How measuring several points on a continuum shows the relation between physical intensity and consumer response, as well as showing which product scores higher, and which scores lower. (b) How measuring just two points on a continuum shows only which product scores higher, but does not reveal the relation between the underlying physical variable and a consumer response.

from product to product, and within a product, the pattern is still evident. Having many more assessors evaluate the products would reduce the variability but would not change the nature of the pattern, nor would it shift the optimal level to a higher or lower level. In contrast, consider the results in Figure 5.3(b), which shows the mean ratings for two samples of the salsa (obtained under a similar test design), but without systematic variation. We no longer look for a pattern. Rather, the investigator then attempts to determine whether one sample scores significantly higher than the other on a specific attribute. Without systematic variation the panel size makes a difference – not to the average rating (which we do not expect to change with increasing panel size), but to the variability (viz. the standard error of the mean). The standard error of the mean decreases with increasing base size. There is a greater likelihood of finding a given difference between products to be 'statistically significant' with greater base size. (It is the discovery of significant differences that is the goal of the study with two products, rather than the location of an optimal point on a liking scale.)

Base sizes of 30–50 per prototype usually suffice in a designed experiment to generate solid, reliable results that can be used to create models relating independent variables to responses. By the time a base size of 50 ratings per sample is reached the mean is well determined. It would be hard to budge that mean with random data, unless a parallel set of 50 ratings is obtained, all of which are random. (This is unlikely.)

There are exceptions, however, to this recommended base size of 30–50 assessors:

1. If there is reason to suspect the existence of different groups of consumers in the population defined by different likes and dislikes, then the investigator should increase the panel size to ensure that the test comprises assessors from the different groups of consumers. As a conservative rule of thumb, it is safe to sample a minimum of 30 assessors from each identified subgroup of consumers. For instance, if the project encompasses responses from men and women, and if one believes that the women's responses will differ from that of men, then it is probably a good idea to sample responses of 30 men and 30 women for each prototype.

2. If the assessor evaluates relatively few products from the test set, it is a good idea to increase the sample size behind each product. Individuals in a test introduce random variability. This variability can be partially cancelled out if each individual in the study tests all or a significant proportion of the samples. However, if the individual only tests one or two samples out of the full set, then the random variability will mask differences between samples. The prudent approach is to increase the base size of the panel, in the hope that the random variability from assessor to assessor will cancel itself out with the larger size.

3. If the method of preparation varies from assessor to assessor, then a larger base size is recommended. For instance, many optimization studies are conducted with products that are used as ingredients, or as accompaniments. These products are mixed in varying proportions with other ingredients. Consequently, inter-individual variability is large, and must be cancelled out by sampling many different individuals.

In a world replete with insecurities it is tempting to over-sample, and expend a great deal of money to ensure that an applied research study truly encompasses the full target population. Researchers often err by working with panel base sizes that are too large. Even though large base sizes ensure better sampling of the consumer population and reduce the variability around the mean, in actuality these large base sizes often cost more than they are worth. Beyond a certain point, and once the researcher has properly sampled the full range of consumer segments in the population, increasing panel size increases cost but does not strengthen the experiment. The relation between physical variables and subjective responses are usually tightly established with relatively fewer assessors. That relation will not change noticeably with increasing base size.

5.2.4   Analysis of the data – a multi-step process

Product optimization studies generate large amounts of data. The data comprise attribute ratings by many assessors of many products (including both systematically varied prototypes as well as benchmark data). The analysis of such data follows a sequence of well disciplined and defined steps, each designed to provide information about the product, and its relation to the physical variables. Furthermore, the richness of the database and the inter-connection among ratings lend themselves to analysis of the relation between sensory characteristics and overall liking, and the relation between overall liking and attribute liking. Finally, the same database can be used to divide consumers into homogeneous groups, based upon how each of the consumers processes the sensory information to generate ratings of overall liking.

The standard analytic steps for product optimization are:

5.2.4.1   Database.   Creation of a database, showing mean ratings of products on attributes. Interest here focuses on the average data. This first step is preparatory.

5.2.4.2   R–R and S–R analysis.   The researcher can proceed along two different paths, R–R analysis and S–R analysis.

R–R (Response–Response) analysis looks at the relation between ratings of different attributes. R–R analysis may focus on understanding consumer language. (Although assessors use many attributes to rate the characteristics of products perhaps they only use a limited number of perceptual dimensions to make their ratings.) R–R analysis may focus on the sensory attributes which drive overall liking or attribute liking. Finally, R–R analysis can deal with the raw data, to uncover clusters of assessors showing similar reactions to products (sensory segmentation).

S–R (Stimulus–Response) analysis refers to the relation between independent physical variables (under the researcher's control), and the dependent variable (consumer ratings). S–R analysis creates relatively simple mathematical equations relating formula variations and responses. The equations comprise linear, quadratic and interaction terms.

5.3   A case history – salsa

The optimization case history shows a specific problem, field work, data analysis, and implications. The concreteness of the problem instructs far better than just a theoretical discussion of available options.

The case history here concerns the development of a spicy Mexican-style salsa. The project involved assessors familiar with salsa, who participated in an extended taste test in which they evaluated 10 of 17 products in a 4½ hour session. Moskowitz (1985) has presented details of the field execution of similar studies.

5.3.1   Experimental design

The experimental design comprised three physical variables (vegetable level, pepper level, hot spice level). With three physical variables one can use a variety of experimental designs, as shown in Tables 5.3, 5.4a and 5.4b. In this study, where interest was focused on liking ratings, it was vital to use an experimental design comprising at least three levels of each ingredient. Thus, the central composite design, shown in Table 5.4b, was the appropriate design.

In addition to the experimental products, the research design investigated two additional products, both currently on the market. All of the products were tested 'blind' (unidentified as to brand or physical composition).

The study was run in three separate markets in the US: Atlanta, Georgia; Chicago, Illinois and Portland, Oregon. The regional sampling ensured (as well as possible) that the study would sample consumers with

Table 5.3 Two level design/three variables

Product	A	B	C
101	1	1	1
102	1	1	−1
103	1	−1	1
104	1	−1	−1
105	−1	1	1
106	−1	1	−1
107	−1	−1	1
108	−1	−1	−1

High = 1, medium = 0, low = −1.

Table 5.4a Three level design/three variables

Product	A	B	C
201	1	1	1
202	1	1	0
203	1	1	-1
204	1	0	1
205	1	0	0
206	1	0	-1
207	1	-1	1
208	1	-1	0
209	1	-1	-1
210	0	1	1
211	0	1	0
212	0	1	-1
213	0	0	1
214	0	0	0
215	0	0	-1
216	0	-1	1
217	0	-1	0
218	0	-1	-1
219	-1	1	1
220	-1	1	0
221	-1	1	-1
222	-1	0	1
223	-1	0	0
224	-1	0	-1
225	-1	-1	1
226	-1	-1	0
227	-1	-1	-1

High = 1, medium = 0, low = -1.

Table 5.4b Three level design/three variables, central composite design (15 of 27)

Product	A	B	C
301	1	1	1
302	1	1	-1
303	1	-1	1
304	1	-1	-1
305	-1	1	1
306	-1	1	-1
307	-1	-1	1
308	-1	-1	-1
309	1	0	0
310	-1	0	0
311	0	1	0
312	0	-1	0
313	0	0	1
314	0	0	-1
315	0	0	0

High = 1, medium = 0, low = -1.

representative taste preferences. (Parenthetically, in early stage developmental projects of this type, the researcher often tests assessors in convenient markets, rather than in many different markets. It is more important at the early development stage to secure cost-effective responses to the different products than to ensure sampling of a product in many markets. With a limited budget, and with 17 products to test, three markets (or sometimes four) is a maximum while still maintaining reasonable costs.)

5.3.2 Results

Table 5.5 shows basic results – the product × attribute matrix. Numbers in the table are average ratings across assessors. These data are available for each of the products, and provide a 'snapshot' of the experimentally varied samples along with the competitors.

5.3.3 Analysis phase 1, R–R analysis

The first analysis uncovers relations among the response variables. R–R analysis can be done profitably for any data set comprising an assortment of similar products. (The products need not be systematically varied.

Table 5.5 Example of database (salsa)

Formulation	110	102	105	106	116	117
Vegetable pieces	1	3	1	1	Comp	Comp
Pepper	2	3	3	3	A	B
Spice level	2	1	3	1		
Overall						
L/D Overall liking	70	66	65	64	47	47
Appearance						
L/D Appearance overall	74	78	72	68	67	64
Sen Light/dark	71	69	76	63	93	58
Aroma						
L/D Aroma	62	62	64	64	65	65
Sen Strength aroma	59	59	63	57	68	65
Taste/flavor						
L/D Taste/flavor	69	68	65	64	52	54
L/D sweetness	61	56	57	57	55	57
L/D Hot flavor	61	63	62	60	56	56
Sen Strength taste/flavor	70	73	79	68	70	76
Sen Sweet flavor	60	53	52	54	55	58
Sen Hot flavor	54	65	68	59	60	64
Texture						
L/D Texture	70	75	75	70	56	58
Sen Thin/thick	75	78	77	69	67	64

L/D = (0 = hate, 100 = love), Sen = sensory, (0 = none, 100 = extreme amount).

Indeed, quite often, a great deal of insight can be gained with an R–R analysis performed on competitive products in the same category purchased off the shelf.)

5.3.3.1   Issue 1 – What is the range of ratings across products?   This first analysis looks at the different attributes in terms of which attributes best differentiate the products. The analysis is performed on the mean ratings, although it could be performed on the raw data. The analysis is simple – just compute the range of ratings for a given attribute. Since the scale is the same, the range shows us whether the products distribute widely on the attribute, or cluster. According to Table 5.6, we find that the attributes differ. For this particular experiment, sensory intensity of both darkness and heat shows the greatest range. The overall flavor/intensity shows a narrow range as does sweetness.

The range of ratings is instructive. It shows us in a rough fashion the magnitude of the sensory effects achieved by the variables. The analysis is purely descriptive, and not predictive, however.

5.3.3.2   Which liking attributes co-vary with overall liking?   As processors of sensory information, consumers pay different amounts of attention to sensory inputs when they judge overall liking. For example, changes in the liking of appearance may be less effective in driving overall liking than changes in the liking of flavor.

Table 5.6 R–R Analysis – Range of the attribute scores

	High score	Low score	Range
Overall			
L/D Overall liking	70	44	26
Appearance			
L/D Appearance overall	78	65	13
Sen Light/dark	93	58	35
Aroma			
L/D Aroma	64	48	16
Sen Strength aroma	64	51	13
Taste/flavor			
L/D Taste/flavor	70	45	25
L/D Sweetness	68	54	14
L/D Hot flavor	63	49	14
Sen Strength taste/flavor	76	58	18
Sen Sweet flavor	64	51	13
Sen Hot flavor	68	34	34
Texture			
L/D Texture	77	60	17
Sen Thin/thick	80	55	25

L/D = liking, Sen = sensory.

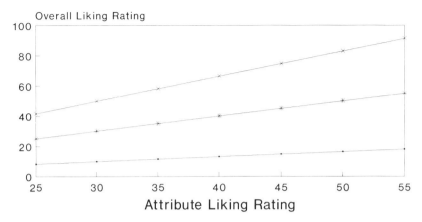

Figure 5.4 Leverage analysis – how unit changes in attribute liking co-vary with changes in overall liking, showing the relative importance of individual attributes. The equations fit to the data follow the linear form: Overall liking = $A + B$ (Attribute liking). High values of B correspond to steep slopes (and more important sensory dimensions): ($\cdot$) slope = 0.33; ($*$) slope = 1.00; ($\times$) slope = 1.66.

The database enables the researcher to measure attribute importance in a straightforward way. The organizing principle is that the relation between attribute liking (independent variable) and overall liking (dependent variable) is typically a straight line. The slope of the straight line measures attribute importance. Consider the prototypical straight lines shown in Figure 5.4. Each has a slope. The higher the slope, the more important the attribute is implied as a 'driver' of overall liking. For instance, a slope of 1.66 means that a 1 unit increase in attribute liking co-varies with a 1.66 unit increase in overall liking. (Small changes in attribute acceptance correspond to large changes in overall acceptance.) In contrast, a slope of 0.33 means that a 1 unit increase in attribute liking co-varies with a 0.33 or ⅓ unit increase in overall liking (viz. a much less important attribute).

Table 5.7 shows the slopes of the lines, and reveals (not unexpectedly) that *taste liking* is the most critical. For comparison, Table 5.7 also shows slopes from a variety of other studies which had 'multi-modal' products, of different appearances, flavors, and textures.

5.3.3.3   How do ingredients and/or sensory attributes drive overall liking?
We know from scientific research that as a sensory attribute increases, liking changes, first increasing, peaking, and then dropping down (Moskowitz, 1981). This organizing principle was discovered more than 100 years ago by the German psychologist Wilhelm Wundt (Beebe-

Table 5.7 Slopes relating attribute liking to overall liking

Attributes	Salsa	BBQ Sauce	Fish	Shrimp
Liking of appearance	0.95	0.92	1.06	0.99
Liking of aroma	1.13	1.03	1.02	1.02
Liking of taste/flavor	1.56	1.67	1.43	1.54
Liking of sweetness	1.15	1.13	–	–
Liking of hot flavor	1.15	1.32	–	–
Liking of texture	1.00	0.89	1.02	1.12

Equation: Overall liking = A + B(Attribute liking).

Center, 1932). Since then, the organizing principle has been shown to operate in a wide variety of cases, with many different stimuli. It is most impressive, however, with taste and flavor stimuli. Figure 5.5 shows a prototypical example of the data. The curve can be based either on variation in stimulus level or on reactions to a more distantly related set of products (e.g. competitor products of the same type in a category, such as tomato sauces).

The organizing principle helps the researcher to understand which sensory attributes in a category are important, and which are not. The prototypical curve shown in Figure 5.5 is really a simple representation of what may be a family of such sensory-liking curves, as shown in Figure 5.6.

The curves may differ from attribute to attribute. For instance, for some sensory attributes consumers may perceive large differences among products (as reflected in the range of the sensory attribute), but this wide range may not co-vary with a wide range of liking. Rather, the relation may be fairly flat. In other cases, the relation may be very steep. In other cases the relation may not be a straight line at all, but rather a curved line (going upwards or downwards).

The curves in Figure 5.6 are 'fitted'. All the points are brought to the curve. Curve fitting to find how sensory attributes drive liking is a qualitative rather than quantitative analysis. We look for general rules.

We can summarize in tabular form how overall liking is driven by sensory attributes. Table 5.8 summarizes the relation between sensory attribute and liking. As can be seen from Table 5.8, the sensory attributes do not generate the same pattern.

5.3.3.4 Do consumers segment in terms of their sensory preferences? A recurrent problem in product development and marketing is the nature of the consumer panel which one chooses to test. We know that people differ from each other. Marketing researchers often link differences in

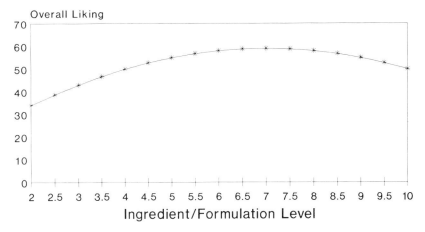

Figure 5.5 How overall liking co-varies with physical ingredient level or sensory level. This is a stylized version of the general inverted U-shaped curve first reported by Wilhelm Wundt (Beebe-Center, 1932): (∗) Total Panel.

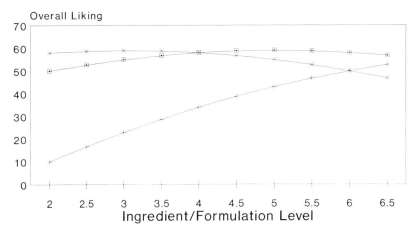

Figure 5.6 How a single curve such as that shown in Figure 5.5 really results from the combination of multiple curves, one from each assessor: (+) assessor 2; (□) assessor 3; (×) assessor 4.

product usage, demographics (age, sex, income), or differences in psychological predisposition (psychographics) to product preferences. In the main these linkages fail. Although there do exist differences among consumers (which are repeatable, and thus reliable), they cannot be traced to standard marketing variables.

Moskowitz (1985) suggested that there exist in the population profound

Table 5.8 How attributes drive overall liking (total panel, Segment A, Segment B)

Attribute	Total	Segment A	Segment B
Darkness	Inverted U	Linear up	Linear down
Aroma	Modest inverted	Linear up	Linear down
Flavor	Inverted U	Linear up	Linear down
Sweetness	Inverted U	Linear down	Linear up
Hotness	Inverted U	Slight up	Inverted U
Thickness	U	Linear down	Linear up

differences among consumers in what they accept versus what they reject. These differences *transcend* the typical market-place variables which marketing researchers investigate. Rather, the population comprises groups of consumers exhibiting fundamentally different sensory preferences. The consumer population can be divided into these fundamental segments, much as any color can be analyzed into three color primaries (blue, yellow, red). Individuals fall into one of these limited set of non-overlapping sensory segments. Market to market differences result from different skews or distributions of these fundamental segments. Sensory segments reveal themselves as different patterns relating overall liking to the same sensory characteristics. The prototypical pattern shown in Figures 5.5 results from combining a limited number of dissimilar patterns, shown schematically in Figures 5.6 and 5.7. Each segment has its own unique pattern relating sensory attributes to liking. The segmentation algorithm develops these individual curves, on a person by person basis, and clusters together individuals with similar curves.

Segmentation pervades responses to products. Salsa products are no different. The algorithm recommended by Moskowitz generated two different segments of consumers. Typical patterns relating liking to sensory characteristics differ dramatically by segment, as Figure 5.7 shows. However, Figure 5.8 also suggests that on some attributes (e.g. appearance, some texture dimensions) the segments agree with each other. (Parenthetically, this means that segmentation will not find dramatic differences on every attribute, just on key attributes.)

5.3.4 Analysis phase 2, S–R (stimulus–response) analysis

Most researchers familiar with product optimization procedures perform S–R analyses, in which they relate the level of the ingredient (or process condition) to subjective responses. Additionally, S–R analysis lends itself to modeling the relation between the physical variables and objective measures (e.g. cost of goods, proximate physical measures, instrumental measures of textural characteristics, etc.).

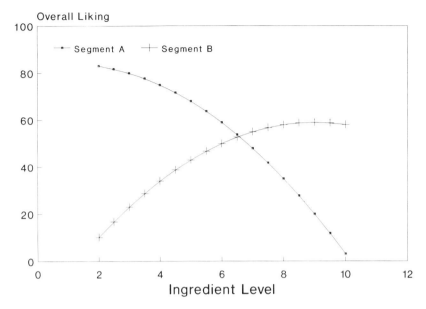

Figure 5.7 Schematic curves showing how two sensory segments, with divergent sensory preferences, respond to the same sensory information in terms of their liking ratings. The sensory segments perceive intensities similarly, but react quite differently in terms of how much they like or dislike what they perceive. There are two key segments. One segment generates curves which slope down with sensory intensity. This segment is labelled Segment A in the data tables, and corresponds to a group of consumers who like 'low impact' salsas (lighter, less spicy, weak flavor 'hit'). The other segment generates curves which increase with sensory intensity. These curves may be steep or flat. This segment is labelled Segment B in the data tables, and corresponds to a group of consumers who like 'high impact' salsas (dark, spicy, strong flavor 'hit').

5.3.4.1 Regression modeling. Data sets comprising independent and dependent variables lend themselves naturally to modeling. Modeling develops an equation which quantitatively relates the levels of the independent variables to the levels of the dependent variables. Modeling uses regression analysis.

Regression equations come in all shapes and sizes. There are some general categories, however, as listed below:

One variable versus multiple variables. Multiple variables as simultaneous predictors enable the researcher to better estimate the likely value of the dependent variable given known levels of the different independent variables.

Types of equations (Linear versus non linear equations). Linear equations assume a straight line relation between the independent and the

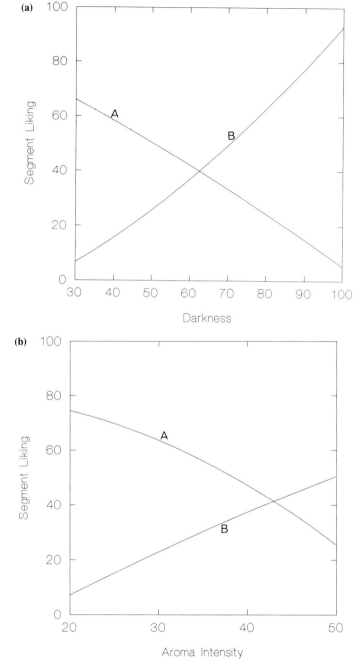

Figure 5.8 Comparison of patterns relating sensory intensity level (abscissa) to overall liking (ordinate). Segment A likes weaker tasting products, and lighter colored ones. Segment B likes more intense products (e.g. hotter tastes, darker colors).

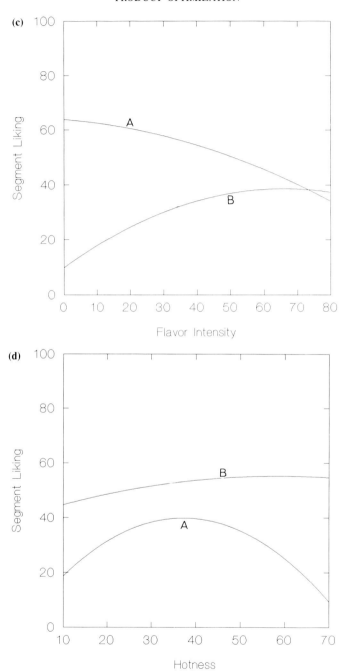

Figure 5.8   (Continued)

dependent variable. A linear equation assumes that unit changes in the independent variable produce the same magnitude of change in the dependent variable, no matter what level the independent variable achieves. Linearity is often violated by liking and sometimes by sensory attributes. The liking relation (versus ingredient level) often follows an inverted U-shaped curve and it is necessary to use a quadratic (non-linear) equation.

Linear equation: Attribute $= A + BX$ (one independent variable, X)
Linear equation: Attribute $= A + BX + CY$ (two independent variables, X, Y)
Quadratic equation: Attribute $= A + BX + CX^2$
Quadratic equation: Attribute $= A + BX + CX^2 + DY + EY^2$

Absence versus presence of interactions. Subjective responses are often determined by interactions among the ingredients. The interactions may be simple (e.g. multiplicative relation) or complex (e.g. ratios). For example in a salsa product, vegetable level and spice level interact. A multiplicative interaction may be modeled as:

$$\text{Subjective rating} = A + B(\text{Veg}) + C(\text{Spice}) + D(\text{Veg}^2) + E(\text{Spice}^2) + F(\text{Veg} \times \text{Spice})$$

The foregoing equation means that the subjective rating will change more dramatically than expected when either the vegetable level or the spice level is high. When vegetable and spice levels are both very high, the multiplicative term will yield an *unexpectedly high* subjective rating.

For other products (e.g. fruit flavored beverages) ingredient interactions may be more complicated. The rating may be related to the *ratio* of two ingredients (e.g. sugar/acid ratio). The rating may be expressed by the equation:

$$\text{Subjective rating} = A + B(\text{Sugar}) + C(\text{Acid}) + D(\text{Sugar}^2) + E(\text{Acid}^2) + F(\text{Sugar/Acid})$$

5.3.4.2 Developing models for the salsa data. The three salsa formulae variables were systematically varied at three levels (Table 5.4b). The design enables the data to be modelled by simple regression. The central composite experimental design allows the investigator to create a variety of different models. The recommended model uses linear and square terms. It allows the cross terms to enter the equation, if the interactions add significant predictability.

Table 5.9 shows the statistics for four models, two relating sensory attributes of 'hot' and the other two relating the liking attribute to the three independent variables. Both linear and quadratic models are shown. The model for 'hotness' is primarily linear. Square terms do not add predictability. The model for 'liking' is quadratic. Square terms add significantly to the ability of the model to describe how liking varies with ingredients. There are no cross terms. (Cross terms in this model would represent significant pairwise interactions.) The cross terms do not appear in the equation because the linear and square terms account for most of the variability in the ratings. (Parenthetically, they usually account for most of the variability in all attributes. Only in very unusual circumstances is an interaction between two variables so strong as to appear after linear and square terms of those variables have already appeared in the equation.)

Equations can be developed for every attribute, whether these be sensory ratings, liking ratings, cost of goods, or instrumental probes. The equations in their totality constitute a *product model*. They enable the

Table 5.9 Linear and quadratic models – parameters of models for overall liking and rated 'hotness'

A. 'Hotness' – linear model

Squared multiple R: 0.924
Adjusted squared multiple R: 0.904
Standard error of estimate: 5.233

Variable constant	Coefficient	T	P (2-tail)
	32.598	5.563	0.000
A	−1.564	−0.945	0.365
B	15.603	9.429	0.000
C	−11.327	−6.716	0.000

B. 'Hotness' – quadratic model

Squared multiple R: 0.947
Adjusted squared multiple R: 0.907
Standard error of estimate: 5.135

Variable constant	Coefficient	T	P (2-tail)
	37.480	2.786	0.024
A	−1.262	−0.099	0.924
B	−5.963	−0.467	0.653
C	3.578	0.269	0.795
A * A	−0.071	−0.022	0.983
B * B	5.450	1.703	0.127
C * C	−3.720	−1.127	0.292

C. 'Overall liking' – linear model

Squared multiple R: 0.384
Adjusted squared multiple R: 0.216
Standard error of estimate: 6.881

Variable constant	Coefficient	T	P (2-tail)
	74.384	9.653	0.000
A	−2.363	−1.085	0.301
B	−1.696	−0.780	0.452
C	−4.957	−2.235	0.047

Table 5.9 *Continued*

D. 'Overall liking' – quadratic model

Squared multiple R: 0.750
Adjusted squared multiple R: 0.563
Standard error of estimate: 5.137

Variable constant	Coefficient	T	P (2-tail)
	59.651	4.433	0.002
A	−27.756	−2.177	0.061
B	37.832	2.961	0.018
C	−1.069	−0.080	0.938
A * A	6.399	2.004	0.080
B * B	−9.983	−3.118	0.014
C * C	−0.988	−0.299	0.772

Squared multiple R ($\times 100$) = Percent of variation in ratings of salsa accounted for by the model.
Adjusted squared multiple R ($\times 100$) = Same as squared multiple R, but takes into account that additional predictors are used.
Standard error of estimate = Standard deviation around the least squares model (a measure of deviation of predictions from empirical data).
Constant = Expected rating, if all variables (A, B, C) held at 0.
T = 'T' test for difference of the coefficient from '0'.
P = Probability that the coefficient is 0 (lower values of P mean that the odds are low that the coefficient is 0, or in other words mean that the statistical term in the model is significant).

researcher to estimate the likely rating (or set of instrumental readings) corresponding to any combination of independent variables *within* the range tested.

5.3.4.3   Using the equations to understand how ingredients drive ratings. One of the simplest analyses holds two of the physical variables constant, and changes one variable in small increments, while estimating the full profile of attribute ratings. This is a 'sensitivity analysis'. It shows the sensitivity of each dependent variable to every independent variable.

Table 5.10 shows sensitivity analysis from the 'center' of the experimental design. Note:

1. An ingredient can influence some attributes but not others.
2. One attribute rating can be affected by several variables but to different degrees.
3. Some attributes are unrelated to formula variables.

Table 5.10 Sensitivity analysis to three ingredients

Ingredient A	1.0	1.3	1.7	2.0	2.3	2.7	3.0
Cost	58	61	63	66	69	71	74
Like/total	66	63	61	60	60	62	65
Like/segment A	59	58	56	55	53	51	49
Like/segment B	39	35	34	34	36	40	46
Like/appearance	66	66	66	68	70	73	77
Like/aroma	53	52	51	51	52	53	54
Like/flavor	65	64	64	64	65	66	68
Like/texture	61	63	65	65	64	64	62
Sensory/dark	72	73	74	75	76	77	78
Sensory/aroma	39	41	42	42	42	40	38
Sensory/flavor	43	46	48	49	49	47	45
Sensory/sweet	14	12	12	11	12	13	14
Sensory/hot	32	34	35	35	35	34	33
Sensory/thick	53	56	58	59	60	61	61
Ingredient B	1.0	1.3	1.7	2.0	2.3	2.7	3.0
Cost	56	59	63	66	69	73	76
Like/total	50	56	59	60	58	54	48
Like/segment A	43	48	52	55	57	59	60
Like/segment B	26	32	34	34	31	25	16
Like/appearance	67	68	68	68	67	67	66
Like/aroma	46	48	50	51	52	52	52
Like/flavor	53	59	63	64	64	60	55
Like/texture	47	56	62	65	65	63	58
Sensory/dark	66	70	73	75	75	75	73
Sensory/aroma	34	38	40	42	43	43	42
Sensory/flavor	26	34	42	49	55	60	65
Sensory/sweet	11	12	12	11	10	7	4
Sensory/hot	22	25	30	35	43	51	61
Sensory/thick	48	53	57	59	60	60	58
Ingredient C	1.0	1.3	1.7	2.0	2.3	2.7	3.0
Cost	54	58	62	66	70	74	78
Like/total	59	59	60	60	60	61	61
Like/segment A	28	37	46	55	64	73	82
Like/segment B	56	48	41	34	28	22	16
Like/appearance	55	59	63	68	72	77	81
Like/aroma	49	49	50	51	52	53	53
Like/flavor	66	65	65	64	64	63	63
Like/texture	82	76	71	65	59	53	48
Sensory/dark	54	61	68	75	81	88	94
Sensory/aroma	49	47	45	42	40	38	35
Sensory/flavor	64	59	54	49	44	38	33

5.3.5   Three examples of optimization technology

This section deals with three separate applications of optimization.

1. *Optimizing*: Maximizing a dependent variable within ingredient limits, either without implicit constraints (e.g. cost of goods), or within pre-designated *implicit constraints* (e.g. within specific costs of goods, or a

product whose sensory profile lies within desired upper and lower limits).

2. *Goal fitting*: Discovering the combination of physical variables, which in concert generate a product whose sensory or image profile is as 'close as possible' to a predesignated profile.

3. *Quality optimizing*: Determining that combination of ingredients which ensures that the maximal percent of product batches in a production run will meet or exceed a quality target.

5.3.5.1   Optimizing a product attribute.   Optimizing uncovers that combination of independent variables which generates a value for the dependent variable as high as possible. The values or levels of the independent variables must lie within the range tested. The product model allows the researcher to explore many alternative combinations of independent variables.

There are two types of optimization – without constraints and with constraints. Optimization without implicit constraints simply finds that particular combination which generates the overall maximum (within the range of variables tested). If we envisage the response surface (or equation) as a 'hill' in many dimensions, then optimization discovers the top of the hill (see Figure 5.9).

The optimum formula level may lie either at the edge of the ingredient range, or in the middle. If the investigator has varied the ingredient levels

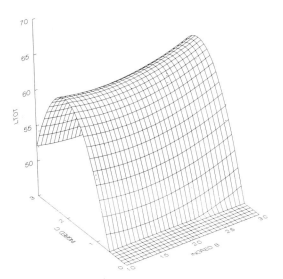

Figure 5.9 How overall liking varies with two ingredients (B, C). The 3-dimensional surface is fitted to the data, and represents the best function (without considering formula variable A). Each attribute will generate its own surface relating formula variables and ratings.

sufficiently broadly, then we expect the optimum level to lie in the middle (especially if the attribute is overall liking). If, as is often the case, the product developer has varied some of the ingredients widely but has been conservative with others, then we may expect to see most of the independent variables optimize at the upper or lower extreme, rather than in the middle.

Table 5.11 shows two unconstrained optimizations for overall liking and sweetness respectively. The overall liking rating peaks at an intermediate level of ingredient B, but at extreme levels of ingredients A and C. For hotness of taste we find another optimal formulation: A = 2.1,

Table 5.11 Sequence of optimized formulations

	L/D Total	Hot taste	L/D Tot Hot > 60	L/D Tot Hot > 55	L/D Tot Hot > 50	L/D Tot Hot > 45
A	1.0	2.1	1.2	1.1	1.0	1.0
B	1.9	3.0	2.9	2.8	2.7	2.5
C	3.0	1.0	1.0	1.1	1.6	1.2
Cost	69	65	57	55	60	53
Liking						
Total	67	47	53	57	59	62
Segment A	85	32	37	38	52	41
Segment B	21	38	42	46	37	50
Appearance	79	53	51	52	59	55
Aroma	55	50	51	51	53	52
Flavor	63	56	58	60	61	64
Texture	43	75	74	75	66	75
Sensory						
Darkness	91	53	51	52	63	56
Aroma	31	49	47	47	43	45
Flavor	25	80	75	71	61	63
Sweet	12	8	11	12	11	15
Hot	27	65	60	55	50	45
Thickness	55	56	52	52	53	53
Instrumental						
Probe 1	28	40	29	27	28	23
Probe 2	−6	655	604	580	407	510
Probe 3	48	49	61	61	60	60
Probe 4	21	46	43	40	37	34
Probe 5	31	45	42	42	39	41
Expert						
Exp 1	20	20	19	19	19	19
Exp 2	41	53	50	49	48	48
Exp 3	1	87	80	77	50	66
Exp 4	2	86	82	78	58	69
Exp 5	13	31	29	28	25	23

B = 3.0 and C = 1.0. (Note that the ingredients range from a low of 1 to a high of 3, so 1 and 3 are the ends of the ingredient range.)

Constrained optimization. A more complicated problem arises when other attributes lie within specific levels. One can wish to constrain the upper and lower limits of one or more sensory attributes. The optimization algorithm searches for an optimal (and viable) combination of ingredients. As more constraints are added (which the optimization algorithm must satisfy), the range of ingredients satisfying the constraints shrinks. (In some cases the constraints are mutually exclusive, so that there is no viable combination of ingredients.)

As examples of the constrained optimization, consider the results shown in the last four columns of Table 5.11. These columns show a series of 'hotness' reductions. Sensory hotness is constrained to go from 0 to high levels. As the highest allowable 'hotness' is systematically reduced from 65 to 45 Table 5.11 shows that the ingredient levels change, as does overall liking.

Prediction of other attribute rating. Product modeling also allows the investigator to estimate the likely profile of attribute ratings that would be assigned to the optimal (or, in fact to any other predesignated combination of independent variables). Table 5.11 shows the expected attribute profile for the optimal products.

5.3.5.2 Goal fitting. One of the properties of the product model is that for a given set of ingredient or processing levels the investigator can estimate the likely set of ratings. One can reverse that logic, set a profile of ratings as 'targets', and search for the combination of ingredient levels that would have generated that set of target levels. In most real world applications it will be impossible to satisfy every goal, but it should be possible to 'come close'.

The algorithm for 'goal fitting' is straightforward. It simply varies the independent variables in a systematic fashion to minimize the percentage difference between target and estimated levels.

Table 5.12 shows results from applying the algorithm. The target profiles can be either sensory attributes, or image attributes. The algorithm first finds the formulation which defines a target goal profile. Then it estimates all other ratings given the formula levels. For instance, the researcher may wish to estimate the expected liking corresponding to a specified sensory attribute level. (This sensory attribute may belong to a competitor product in the same category.)

The goal-fitting approach enables the researcher to estimate the likely set of ingredient levels (within the experimental design) that would have generated the target sensory profile. Once the set of ingredient levels is

Table 5.12 Formulations delivering a desired 'goal or target' profile

	Consumer goals				Instrumental goals			
Formula								
A	1.01		3.00		1.89		2.13	
B	1.00		1.49		1.11		1.10	
C	1.34		1.00		2.95		2.99	
Cost	40		57		68		70	
Liking								
Total	55		62		53		53	
Segment A	29		16		71		71	
Segment B	45		67		11		11	
Appearance	56		64		79		81	
Aroma	47		50		49		49	
Flavor	54		66		54		54	
Texture	55		73		34		33	
	Act.	Goal	Act.	Goal				
Consumer sensory								
Darkness	49	45	54	30	85		87	
Aroma	53	25	41	25	28		28	
Flavor	30	30	49	50	14		13	
Sweetness	16	15	20	20	9		9	
Hotness	20	30	28	50	19		19	
Thickness	40	40	54	60	51		52	
					Act.	Goal	Act.	Goal
Instrument probes								
Probe 1	7		33		29	30	32	50
Probe 2	400		629		5	5	7	8
Probe 3	53		42		32	30	30	20
Probe 4	7		19		9	10	8	20
Probe 5	28		39		27	25	27	40
Expert panel rating								
Exp 1	15		18		19		19	
Exp 2	34		43		35		35	
Exp 3	50		83		2		3	
Exp 4	54		79		1		0	
Exp 5	3		12		3		3	

Act. = Actual (estimated by the model).
Goal = Set as goal by experimenter.

determined, the researcher can straightforwardly estimate the expected liking rating. Alternatively, the researcher can obtain 'image' attributes for the product assigned by consumers, and relate them to ingredients. Then, in the goal fitting portion, the researcher assigns a target 'image profile' to the product, and discovers the likely ingredient levels that would yield the image profile, as well as predict overall liking for this combination. In this way it is possible to link the non-concrete 'image' attributes directly to ingredients.

Relating two diverse data sets by goal fitting approach. Relating different sources of data has always been of interest to researchers. For instance, the researcher may collect data from instruments, using the same prototypes as the consumers evaluate. The objective is to estimate the consumer reactions given the profile of measures provided by the instrument probes. A probe might be an instrumental measure of texture, a chemical or an optical analysis.

To interrelate two sets of data using optimization technology requires that the investigator obtain physical measures on the same set of products. These physical measures are treated as 'attributes' in the same way that the consumer ratings are treated as attributes. Regression modeling yields a set of equations relating the independent variables to the physical measures.

The same logic can be extended to other dependent variables, such as ratings by experts or ratings by quality control panelists. In each case the data for the other 'set' of variables (experts, instruments, quality control) is obtained from the same set of stimuli. The equations are developed by standard regression analysis. The result is an augmented scale product model, interrelating physical variables under the researcher's control to consumer data, instrumental data, ratings by experts, ratings by quality control personnel, and cost of goods.

To illustrate the application, consider the results in Table 5.12. There were five physical 'sensors' or probes which generated instrumental measures of the products. These generated five equations, one per sensor. The results in Table 5.12 show the estimated formula level corresponding to the specific profile of the probes obtained from a new sample. Table 5.12 also shows the estimated liking rating as well as the estimated profile of sensory attributes.

Goal fitting provides the researcher with a mechanism to interrelate different sets of data. Over the years, researchers have correlated data from two sources to determine whether these data sets are related to each other. Standard analytic procedures include regression analysis, whereby the researcher relates a dependent variable (e.g. consumer acceptance or perceived sensory attributes) to one or several independent variables (e.g. instrumental measures). Traditional research has not, in general, provided useful predictions of the magnitude of the dependent variable from the levels of the independent variables. Research usually stops after demonstrating a significant relation.

Goal fitting techniques enable the investigator to predict the likely profile of attributes in one set of data from values in another set. The method is potentially very powerful for product development and quality engineering. Product developers can work with internal sensory panels or instruments, to predict the likely profile of consumer reactions (e.g. acceptance, sensory attributes) from expert panel ratings. Quality assurance engineers can use the technology to estimate consumer acceptance for products whose physical measures are available from sensors. The quality control engineer can obtain measures on a batch by batch basis, and use these measures to estimate the likely consumer acceptance level corresponding to the batch.

5.3.5.3 Quality modeling and production variability. This application deals with the variability of production, rather than with a single formulation that maximizes a criterion variable. With salsa varying on three physical ingredients simultaneously we expect batch to batch variability. Despite a single target formulation for production, the 'real world' of production will generate a distribution of products, with different ingredients. The variability of the ingredients will contribute to the variability of the ultimate product.

Previously we dealt with estimates of single 'points' or formulations satisfying one or several objectives (e.g. maximize overall liking to generate a product whose sensory profiles matches a target). Here we deal with variability of products – specifically, the variability of overall liking and/or the variability of sensory characteristics.

The key objective of quality modeling is to determine how production variability affects consumer acceptance and the sensory integrity of the product. A further objective is to determine the range of production variability (e.g. from batch to batch) for which variations in product acceptance and sensory characteristics are acceptable. It may be impossible to reduce production variability to 0, but there should be some acceptable region of production variability.

The quality modeling approach comprises six steps:

Step 1. Designed experiment. The designed experiment consists of a systematic array of production levels and production variability. Table 5.13 shows one example of a design, comprising 29 combinations. There are three formulation levels for each of the ingredients, and three standard deviations, one for each ingredient. The formulation levels correspond to the production specification. The standard deviations correspond to production variability. The experimental design combines these six variables so that they are independent of each other. (The design is a 1/4 replicate, central composite design.) Each line in Table 5.13 corresponds to a possible production scenario – viz. specification levels for formulation, and expected variabilities of production around that specification.

Step 2. Production batch simulation. Each row in Table 5.13 can be 'expanded' to a distribution of 100 production batches. The mean of the batches will be the mean ingredient level shown in Table 5.13. However, each ingredient will vary in those 100 batches, with the standard deviation

Table 5.13 Experimental design used to develop quality statistics

Product	Target level for production			SD		
	A	B	C	D	E	F
101	3	3	3	0.3	0.3	0.3
102	3	1	3	0.1	0.1	0.3
103	3	1	1	0.3	0.3	0.3
104	1	3	3	0.3	0.1	0.3
105	3	3	1	0.1	0.1	0.3
106	1	3	1	0.1	0.3	0.3
107	1	1	1	0.3	0.1	0.3
108	1	1	3	0.1	0.3	0.3
109	3	1	1	0.1	0.3	0.1
110	3	3	3	0.1	0.3	0.1
111	1	3	1	0.3	0.3	0.1
112	3	1	3	0.3	0.1	0.1
113	1	1	3	0.3	0.3	0.1
114	3	3	1	0.3	0.1	0.1
115	1	3	3	0.1	0.1	0.1
116	1	1	1	0.1	0.1	0.1
117	3	2	2	0.2	0.2	0.2
118	1	2	2	0.2	0.2	0.2
119	2	3	2	0.2	0.2	0.2
120	2	1	2	0.2	0.2	0.2
121	2	2	3	0.2	0.2	0.2
122	2	2	1	0.2	0.2	0.2
123	2	2	2	0.3	0.2	0.2
124	2	2	2	0.1	0.2	0.2
125	2	2	2	0.2	0.3	0.2
126	2	2	2	0.2	0.1	0.2
127	2	2	2	0.2	0.2	0.3
128	2	2	2	0.2	0.2	0.1
129	2	2	2	0.2	0.2	0.2

SD = Standard deviation of production.

of the 100 batches being the standard deviation shown. Thus, each row in Table 5.13 corresponds to a production scenario. If we were to calculate the means, they would be close to the means shown in Table 5.13. The standard deviations would be close to the standard deviations shown in Table 5.13.

Step 3. The expected variability of consumer ratings. Every line in Table 5.13 (a production scenario) can be expanded to simulate a set of three ingredients. Each batch generates its own sensory and liking profile (based upon the model relating ingredients to consumer attribute ratings). For the 100 simulated batches, we compute the variability of the attribute rating, for each attribute separately. (In practice we compute the standard deviation of each attribute.)

Step 4. Relating consumer attribute variation to ingredient level and production variability. The experimental design in Table 5.13 enables the researcher to model the standard deviation of each attribute as a function of the ingredient level and the production variability (viz. ingredient standard deviation). The model is a familiar one – viz. comprising linear and square terms. The independent variables are the ingredients and their standard deviations. The dependent variables are now the *standard deviations* of each consumer attribute. (Other quality statistics could be modeled as well, such as the percent of batches in each production scenario that achieve or exceed a given liking rating.)

Step 5. Learning from the quality model. Table 5.14 shows how the standard deviation of a selected group of consumer attributes changes with variations in the production variability of each ingredient. The sensitivity table was developed from the 'midpoint', with the following production scenario:

Each ingredient based at 2 (middle of the ingredient range).
Each ingredient had a production standard deviation of 0.2.

Table 5.14 Variability of consumer attributes as a function of ingredient variability in production

SD A	0.5	0.10	0.15	0.20	0.25	0.30
Like/Appearance	11	12	12	13	13	13
Sensory/Dark	0	2	4	6	8	11
Sensory/Aroma	4	6	8	11	13	15
Sensory/Flavor	21	24	31	37	44	50
Sensory/Hot	18	26	34	41	49	46
SD B	**0.05**	**0.10**	**0.15**	**0.20**	**0.25**	**0.30**
Like/Appearance	10	10	9	9	8	8
Sensory/Dark	0	1	3	4	6	7
Sensory/Aroma	1	1	1	1	1	1
Sensory/Flavor	11	11	10	10	10	10
Sensory/Hot	10	9	8	7	6	5
SD C	**0.05**	**0.10**	**0.15**	**0.20**	**0.25**	**0.30**
Like/Appearance	16	21	25	30	35	40
Sensory/Dark	9	20	31	42	52	63
Sensory/Aroma	4	6	8	10	12	14
Sensory/Flavor	16	20	25	29	34	38
Sensory/Hot	10	9	9	8	8	7

Interpreting the table:
Simulate 100 batches of production, with a target ingredient level of 2 units for ingredients A, B and C, respectively.
Look at the distribution of attribute ratings (e.g. sensory darkness) for these 100 batches.
Hold the variability of each ingredient at 0, except the one ingredient in bold.
SD = Standard deviation.

The sensitivity analysis was done in a manner parallel to that in Table 5.10. Each ingredient or standard deviation was held constant at the midpoint. The production standard deviation of any ingredient was then varied in small increments over the range, and the standard deviations of the consumer attribute estimated.

Table 5.14 shows that changes in production variability can substantially change product integrity. The effects differ, however. Increasing ingredient variability may in some cases *decrease* attribute variability, possibly for two reasons. One reason is that the ingredient 'masks' or 'suppresses' an attribute. Increasing production variability of that masking agent may diminish the variation of the masked attribute. The second reason is that increased production variability of an ingredient shifts much of the ingredient range to an area where the sensory attribute is no longer sensitive to changes in the ingredient (viz. region where the sensory intensity asymptotes with ingredient level). For the most part, however, increasing production variability of ingredients increases the variability of a sensory attribute, thus decreasing sensory integrity. Furthermore, liking variability also increases as well.

Step 6. Using the quality model to discover feasible production variability. Table 5.14 shows us regions of different sensory integrity. If our criterion measure is identity or reproducibility of 'appearance' (as indexed by liking of appearance), we find that we need not control Ingredient A. Changes in production variability of Ingredient A exert little effect on the variability of 'liking of appearance'. Removing controls over Ingredient A will change the darkness of the product, so that the variability of dark color will increase as A fails to be controlled. However, that will not change how much consumers like what they see. Ingredient B follows the same pattern. Ingredient C is critical, however. Removing production control will both result in wild swings of perceived darkness, and of liking of appearance.

The quality model can also be used to determine what combination of production standard deviations need to be targeted, to achieve a given range of product sensory integrity. That is, if the researcher can identify those attributes which are critical to maintain (viz. where sensory integrity is critical), it becomes possible by goal fitting to discover the production variability (viz. ingredient standard deviations) generating that profile of sensory integrity.

The quality model described here represents a further step in optimization technology. The quality model is one step removed from the consumer data. It requires simulation of production not just formulation.

5.4   An overview

Product optimization is fast becoming one of the product developer's key tools to increase productivity. The ground work of the technology was laid decades ago. It is only with the advent of the personal computer, however, that optimization has truly grown into a useful tool. Product developers and marketers now use the technology on a regular basis to learn about the dynamics of products, and to improve both overall product acceptance and product sensory characteristics.

As this chapter shows, the optimization technology has not simply remained a statistical procedure to be used blindly. Like other areas of statistics, optimization has taken on a life of its own. The method itself is straightforward, the statistics are fairly simple, and the algorithms widely known and understood. What is fascinating to observe, however, is how the tools are transformed by users into a variety of applications, ranging from sensitivity analysis, to optimization, to goal fitting by reverse engineering, to quality control. The applications truly give the optimization technology the forum and opportunity it needs to display the power which resides within it.

We may, in the next decade, expect to see many new applications of the technology, as the computer makes it possible to implement optimization in a real world environment. Some possible technologies (*vis-à-vis* consumers) are:

1. *Process quality control.* Sensors can be plugged into the processing function, to take instrumental readings of batches. These readings, in turn, can be used to estimate the likely consumer reaction to the products. Reverse engineering will lead to improved process control, with machines programmed to act as human beings.
2. *Databasing.* The technology will allow an in-house panel of experts to act as surrogate consumers. Their ratings of competitor products, along with the optimization technology, will enable manufacturers to copy competitors, using the manufacturer's own ingredient systems.
3. *Computer-aided product development.* Experimental design is used to create prototypes in a rapid, evolutionary way. Machines will be hooked up to control the dispensing and processing of ingredients. These machines will create the systematic product variations, make the physical measures, and then modify the physical inputs until the resulting product generates an instrumental profile which predicts maximal consumer acceptance. At this point product development will have become truly computerized, with much of the tedium removed. Machines, under control programs, interfaced with the optimization technology, will become creators of prototypes, very similar to the way that computer-aided design enables engineers to create prototypes by

computer control of drills and lathes. Product developers will be freed from the tedium, and finally empowered to use their innate creativity.

References

Baxter, N.E. (1989) Research guidance: Not giving it your 'best shot', in L. Wu (ed.) *Product Testing With Consumers*, American Society For Testing And Materials, STP 1035, Philadelphia, pp. 10–22.

Beebe-Center, J.G. (1932) *The Psychology Of Pleasantness And Unpleasantness*, Van Nostrand Reinhold, New York.

Box, G.E.P., Hunter, W.G., and Hunter, J.S. (1978) *Statistics For Experimenters*, John Wiley, New York.

Gordon, J. (1965) Evaluating sugar-acid-sweetness relationships in orange juice by a response surface approach. *Journal Of Food Science*, **30**, 903–7.

Meiselman, H.L. (1978) Scales for measuring food preference, in M.S. Petersen and A.H. Johnson (eds) *Encyclopedia of Food Technology*, Avi, Westport, CT, pp. 675–678.

Moskowitz, H.R. (1981) Sensory intensity versus hedonic functions: Classical psychophysical approaches. *Journal Of Food Quality*, **5**, 109–38.

Moskowitz, H.R. (1985) *New Directions For Product Testing And Sensory Evaluation of Food*, Food And Nutrition Press, Westport, CT.

Moskowitz, H.R., Jacobs, B.E., and Firtle, N.M. (1980) Discrimination testing and product decisions. *Journal of Marketing Research*, **17**, 84–90.

Stevens, S.S. (1975) *Psychophysics: An Introduction to its Perceptual, Neural And Social Prospects*, John Wiley, New York.

Yoshida, M. (1964) Studies in the psychometric classification of odor. *Japanese Psychological Research*, **6**, 111, 124–55.

6 Preference mapping in practice
K. GREENHOFF and H.J.H. MacFIE

6.1 Introduction

Consumer research involving the assessment of products, be they food, beverages, fragrances or household products, traditionally takes the form of the paired preference test, preference ranking or hedonic scaling, usually on two, but sometimes three or four products. The tests are generally easy to conduct, easy to analyse, and are generally thought to give a good measure of relative acceptance or product preference. However this type of research does suffer from several disadvantages, the most important of which is that it can be very limited in providing clear diagnostic information about why a product performs the way that it does. The reasons for this are several-fold:

- Consumers have a very limited vocabulary when it comes to describing their perceptions of products.
- They often use attribute scales incorrectly and are subject to various biases when completing questionnaires.
- Interpretation of paired test data can be more difficult than it first appears.

Most consumers, even the most sensitive, are generally unable to describe objectively what it is that they are perceiving. Given the chance to record their likes and dislikes about products, one will generally find that the majority of their comments will be hedonically based rather than related to specific product attributes. For example, beverages that are liked will usually be described as 'nice', 'pleasant', 'tasty', 'fresh', 're-freshing', 'clean tasting', whereas those that are disliked will receive similar negative responses. In truth consumers do tend to be more de-scriptive about what they dislike rather than what they like with responses like 'too harsh', 'too chemical', 'too synthetic', 'too much bite', 'too tangy'. But even in this situation it can be difficult to infer the actual sensory basis of their criticisms. Researchers therefore try to overcome this natural limitation in consumer vocabulary by presenting consumers with a series of attribute scales for them to rate. Ignoring the debate about whether the types of scale chosen are appropriate or not, one generally assumes that all respondents will understand what we mean by a

particular attribute (i.e. bitter, firmness, fragrant, etc.) and that all consumers interpret the term in a similar way. This is clearly a false assumption. Even trained members of sensory panels commonly confuse
terms such as bitterness, astringency, sourness, so what hope for the
consumer? Also if one asks the consumer to judge the acceptability of a
product and subsequently rate it on a series of product attributes, there is
a tendency for many of the consumers to justify their hedonic response in
terms of the product attributes; thus a consumer who says that a product
is very acceptable will be unlikely to give a negative response to specific
attributes, whereas someone who finds a product unacceptable will tend
to be critical on every attribute. This seems to be a good way to discover
which attributes consumers think are positive product characteristics and
which are negative, but in terms of interpreting consumer acceptance it
can be less than ideal because individual attributes have not been rated
objectively.

The interpretation of paired test data is often difficult. If products
differ on more than one characteristic (and they usually do), how does
one decide which of several differences are important, and in particular
which attribute is crucial in deciding preference? The reality is that the
researcher makes an educated guess, or alternatively suggests a series of
paired tests in which each potentially important attribute is modified and
tested against the original, a fairly wasteful procedure in terms of both
cost and interpretation.

Another problem with univariate analyses of this type of data is the
implicit assumption that all subjects exhibit the same behaviour, and that
a single mean value or some other summary statistic is representative
of all the subjects. One solution is to inspect the distribution of data
for multimodality, and to conduct separate analyses on the various sub-
groups. However, a very large amount of data is usually required for
separate modes to become apparent, and even then may not reveal
significant differences between definable respondent sub-groups.

6.1.1 An alternative approach

Multidimensional preference mapping is a class of alternative methods
that overcome these disadvantages. The basic data is collected by requiring consumers to assess a larger number of products (6 or more), but
assessing the products hedonically (scaled acceptance, rank preference, or
alternatively suitability for purpose, appropriateness to context, etc.).
Then, unlike conventional analysis, individual subject differences are not
averaged, but are built into the model and play an integral role in the
fitting algorithm. However there are two distinct ways of dealing with this
data, known as internal and external analysis (Carroll, 1972). With
internal analysis the objective is to achieve a multidimensional representa-

tion of the stimuli based solely on the acceptance/preference data, whilst in *external analysis* the aim is to relate product acceptability to a multi-dimensional representation of stimuli derived by other means (i.e. from sensory or instrumental data).

The aim of this chapter is to give a non-technical introduction to both types of preference mapping, including case studies from our own experience.

6.2   External preference analysis – Prefmap

6.2.1   The method

With this type of analysis the acceptance/preference data of a group of subjects is related to a multidimensional representation of the stimuli, usually derived from other non-preference data relating to the stimuli. The stimulus space is most usually derived by Principal Component Analysis of descriptive sensory data generated by a trained panel (Piggott and Sharman, 1986). It could just as easily be derived by similarity or dissimilarity estimation in conjunction with multidimensional scaling (Schiffman *et al.*, 1981), or from attitudinal data, or alternatively from physical/chemical measurements.

The principle of external preference mapping is very simple. The acceptance data *for each individual* are regressed against the product co-ordinates obtained from the multivariate analysis of the external data. Regression models may be linear, or involve squared or interactive terms, the latter enabling surfaces to be constructed. The maximum point of a surface may be interpreted as a point of ideal acceptance and indicates the position of the ideal product for that particular individual.

In its original form the external Prefmap algorithm from Carroll (1972) includes four phases of analysis (Figure 6.1). The simplest regression model is a linear or vector model (described in the algorithm as phase IV). This implies a direct relationship between one or more characteristics which are increasing or decreasing across the stimulus space. Thus the more a particular characteristic is present, the more it is liked, or alternatively the less a particular characteristic is present (i.e. an off flavour or taint), the more the products are liked.

A more complex model is the ideal point model. This type of model infers that some stimuli will have too much of a particular characteristic, whilst others will have too little, so that there will exist at some point within the stimulus space an ideal strength of that characteristic, which will equate with an ideal point. Carroll (1972) in fact suggested several variants on the ideal point model. The first possessed circular contours, such that at a fixed distance from the ideal point the acceptance would be

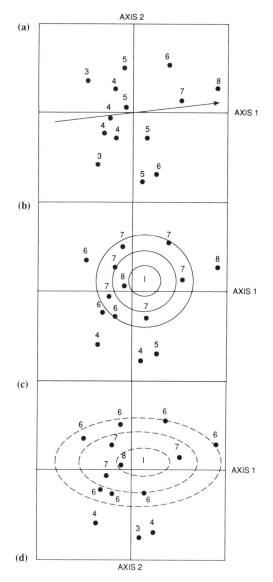

Figure 6.1 Alternative models for preference mapping (external analysis). (a) Vector model (phase IV), (b) circular ideal point (phase III), (c) elliptical ideal point (phase II), (d) phase I is elliptical ideal point rotated between axes 1 and 2.

constant regardless of direction (phase III model). The second possessed elliptical contours instead of circular ones, implying that the dimensions of the stimulus space do not contribute equally to the acceptance model. He further distinguished between elliptical models whose contour axes were parallel to the underlying dimensions (phase II) and elliptical models where the contour axes are rotated to some point between the dimensions (phase I).

All the ideal point models postulate that there is some optimum combination of the external measurements which represent the ideal point for a particular subject. Their acceptance/preference for any stimulus is, therefore, directly related to its similarity to the ideal point. One should emphasise again that this is in direct contrast to the vector model, which assumes a single direction of increasing acceptance/preference throughout the space, but with no accurate measure of the optimum value required for the external characteristics.

In practice the data for all subjects is fitted (regressed) onto the external coordinates for the stimuli using all four Prefmap models. The variance explained by each model is analysed to identify subjects showing a satisfactory fit on particular models. For example, those respondents who are significantly fitted using the vector model and for whom the ideal point models show no improvement, are plotted on the appropriate external dimensions.

Those respondents who are not fitted on the vector model but are fitted significantly on the ideal models, as well as those individuals who are significantly better explained by the ideal point models compared to the vector model, are likewise plotted on the external dimensions.

All remaining respondents who have low levels of variance explained by all four models are concluded to have been unaffected by the variation in the characteristics expressed in the external dimensions, and may be subjected to a secondary internal preference analysis.

6.2.2 Case study using external analysis

A study was carried out on a ready-meal recipe (casserole) in which the only variation intended was in the meat phase. Indeed, the manufacturer wished to replace meat with a non-meat material, but desired to establish how variation in texture might influence product acceptability. Therefore the ready-meal recipe was prepared using 10 variants of the non-meat material. An additional four different cuts of meat were used to provide reference points. Descriptive sensory analysis was carried out by the manufacturer's trained panel with meat/non-meat texture described using nine sensory characteristics. Principal component analysis on the texture data revealed that 86% of the original variance could be explained by just two components, the first of which might best be described as a soft–firm dimension, and the second a dry–juicy/moist dimension. Both meat and non-meat variants demonstrated similar ranges in the dry/juicy continuum but the firmness measurements, although continuous, did tend to separate the meat products from the non-meat variants, the latter generally described as less firm than the meat reference products.

Consumer trials were conducted in two regions of the United Kingdom. Female respondents were pre-recruited by market research interviewers, who established that the respondents fulfilled certain criteria of eligibility, in particular their current purchasing pattern of ready-meal dishes. The respondents then attended tasting sessions at local venues, where they were briefed about the nature of the test (including the number of products) and asked to evaluate the acceptability of the 'meat' phase of the ready-meal using a 9-point hedonic scale. Respondents assessed all the variants presented over the course of a two-hour session, but were expected only to indicate how much they liked the product (using a 9-point category scale), and to describe in their own words what they liked or disliked about the products. Each respondent received the products in a unique order generated by a computer design intended to achieve a complete balance of order and position every 14 respondents (MacFie et al., 1989).

Acceptance data were first subjected to analysis of variance using product and respondent classification as factors; then preference mapping was carried out by regressing each individual's acceptance measures onto the product coordinates obtained by principal component analysis on the sensory data. Only the coordinates from the first two components were used.

194 respondents completed the study generating acceptance data on all 14 products. Examination of the Prefmap output revealed that 94 respondents (48.5%) were significantly fitted ($p = 0.05$ or less) onto the sensory space using the vector model (phase IV). A further 41 respondents (21%) were best explained by one of the ideal point models, leaving 59 respondents (30.5%) who could not be regressed onto the sensory space at a satisfactory level of statistical significance.

Figure 6.2 illustrates the preference vectors of those respondents best explained by the vector model (phase IV). The overall direction of acceptability, established by correlating the total sample mean scores with the product coordinates, is shown to be in the direction of the top right quadrant, indicating that products A and B are most acceptable, followed by C and test variant f, whereas product D and variant i are least acceptable. Since the average acceptance vector is well fitted, one can recover the approximate acceptance ranking from the projection by dropping perpendiculars from each product onto the vector line. Each of the 94 respondents, who are significantly fitted, are shown on the plot as an asterisk. This represents the end-point of each preference vector, which can be visualised as a vector by drawing a line between an asterisk and the centroid. However it is not recommended that the actual vector lines are drawn onto these plots if there are more than about 30 to 40

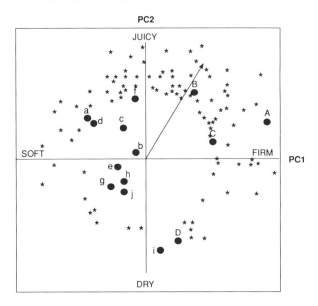

Figure 6.2 Demonstration of external analysis on ready-meals. Significant subject preference vectors (∗) fitted on to the sensory space derived by principal component analysis, PCs 1 and 2 describing 86% of original variance. ● = Product positions within the sensory space, meat reference products coded A to D, non-meat test variants coded a ~j. Direction of overall acceptance (mean scores) shown by →.

subjects; it simply becomes too cluttered. The figure also illustrates a modification to our programme that is not available in the original (Conall) method. In our Genstat implementation of the preference mapping methods, the assessors are not placed around the edge of a unit circle, but placed at a distance corresponding to the variance accounted for by the two dimensions of the plot.

Examination of the preference vectors reveals a considerable diversity of opinion across the sample of respondents, but by relating their positions with the known sensory dimensions within the external data, one can infer the respondent perception which determines their preference choice or acceptability rating. Thus respondents positioned to the right of the plot prefer firmer, less soft products, whereas those to the left prefer less firm products. Similarly, respondents positioned to the top of axis are influenced by moistness, preferring the more juicy/moist products and rejecting drier textures. Respondents situated at the bottom of axis 2 prefer the drier textures. Respondents situated between the two axes are influenced by both dimensions as appropriate. Considering the respondents as a whole, the majority of these fitted by the vector model are reacting positively to moistness, some preferring firmer products and others preferring less firm products. Relatively few respondents would prefer to eat drier products. As a result the manufacturer of the non-meat variants has a clear indication of how to modify his products in order to increase acceptance. Variant f was the best of the non-meat variants. It was most moist, and one of the more firm products. Its acceptance would increase if it were made firmer, perhaps approaching the acceptability of meat B, or if it were made more moist still. However since these respondents were all fitted using the vector model, which implies the more of a particular characteristic the better it will be liked, we cannot accurately define the upper limit to moistness and firmness. We cannot predict how far we can increase these parameters before respondents begin to perceive them as excessive. To establish this we would need to conduct a further study in which both parameters are extended further.

Figure 6.3 illustrates those respondents who are best explained by the ideal point models. Since as a group they represented just 21% of all respondents, further segmentation into respondents best fitted by phase III (circular) or phase II (elliptical) has not been carried out in this case. Interestingly most of these respondents are quite closely clustered together in the sensory space between variant b and product C. Thus one would infer that these respondents find the firmness of the test variants somewhat lacking, whereas the meat products are too firm. In terms of

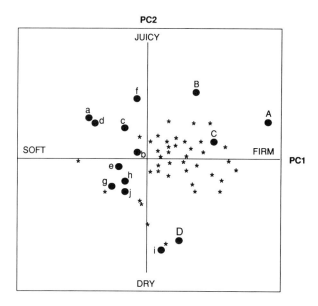

Figure 6.3 Demonstration of external analysis on ready-meals. Significant subject ideal points (∗) fitted on to the sensory space derived by principal component analysis.

moistness, one would specify a level close to the average as being most suitable for the majority of these respondents.

The conclusion of the external analysis shown is that one can separate respondents into discrete segments defined by similarity of response; one can quantify each segment, identify it in terms of socio-demographic details, attitudes, or behaviour depending upon the information collected; and one can produce a sensory specification for any preference segment that is considered interesting or worthy of further development.

6.2.3 Problems with external preference mapping

What of the 30% of respondents who are not significantly explained by regression onto the external data? We have previously stated that these respondents have not reacted to the variation in texture described by the study, which is probably a slight over-simplification. As a group these respondents could probably be further sub-divided into three types. The smallest group is likely to be those who have not reacted to the sensory variation in the products. There are always some respondents who describe all the products as equally acceptable, or equally unacceptable, regardless of the sensory variation. If all the product scores are the same, regression is clearly not possible. Another group will consist of those respondents who are not interested in the main sources of sensory variation, but will have reacted to other sensory differences. For example, an external configuration of stimuli generated from principal component analysis accounted for 86% of the original variance in two components, with 14% of the original variance still unaccounted for. It is always possible that apparently minor sources of sensory variation are important to some individuals, and these will be missed if further components are not included in the analysis. Also because the initial objective of this study was to assess the importance of product texture, details pertaining to product appearance or flavour were ignored. It may be that some respondents reacted to differences in appearance or flavour that were not represented by, or related to, the existing external data. One solution, given sufficient stimuli material, would be to extend the sensory assessment retrospectively and re-analyse the non-fitted respondents on this additional external data.

Finally there will always be some respondents who are unable to complete the task required of them. Although describing product acceptance is much easier than describing products on a series of specific characteristics, it does involve translating perception into some form of scale, and unfortunately there will always be a finite number of respondents who are unable to fulfil this task reliably. All forms of consumer research are faced with this problem, and very few attempt to quantify the erratic or unreliable contributors to the data.

6.3   Internal analysis – MDPREF

6.3.1   Method

An alternative procedure which may be used to investigate hedonic data, is to produce a product space based on the acceptance data alone. This is internal analysis or true multidimensional preference analysis (MDPREF). The method is often described using this acronym which was the original algorithm (Chang and Carroll, 1968). It is perhaps simplest to think of this as a variation of principal component analysis. In PCA, a product (columns) × attribute (rows) matrix is reduced to a smaller number of independent components or factors whilst minimising the loss of original information. Each of the original attributes is then projected onto the resulting components in order to interpret them, the cosine of the angle between the attribute vector and each component reflecting the correlation between them, and the length of the vector being proportional to the variance explained. With MDPREF one would replace the attribute data in the previous matrix with acceptance data for each subject. Therefore the matrix becomes products (columns) × subjects (rows). Then in the same way that principal component analysis identifies the major source of variation and extracts this as component 1, MDPREF identifies the major variation within the preference data and extracts this as preference dimension 1. It then proceeds to identify a second preference dimension orthogonal to the first, and so on until all the variance in the acceptance data is explained. In the same way that product attributes can be related to the extracted components from PCA, so the individual respondent acceptance vectors can be related to the extracted preference dimensions, the cosine of the angle between an individual preference vector and the dimension reflecting the correlation between them, and the length of the vector being proportional to the variance explained. For each preference dimension, product coordinates are generated which allow the preference responses of these respondents correlated with that dimension to be recreated, resulting in the construction of a multidimensional product map based on acceptance data alone. A very simple example is shown in Figure 6.4.

This type of preference mapping, like external analysis, requires that all respondents assess all products and generate a score for acceptance or some other measure. The first stage of the analysis is usually to centre the scores of each individual to zero, then to scale them to unit variance. This is followed by the internal preference analysis itself. Output from the analysis includes the cumulative variance accounted for each subject over all extracted dimensions, and graphs of stimuli and subject relative to all pairwise combinations of the first three principal axes. Additional axes could be plotted but since MDPREF extracts dimensions in decreasing

Raw Data Matrix - Acceptance Scores on 5 Products

Assessor	A	B	C	D	E	
A1	1	2	3	4	5	)
A2	5	4	3	2	1	)
A3	1	2	3	4	5	)
A4	3	1	5	1	3	) Acceptance Scores
A5	3	5	1	5	3	)
A6	2	4	4	4	2	)

A)   The acceptance scores for assessor 1 can be represented as a linear preference dimension, with high acceptance towards E.

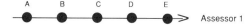

B)   The acceptance scores for assessors 2 and 3 could be similarly represented using this dimension, subject only to the observation that assessor 2 exhibits the opposite behaviour to assessors 1 and 3

C)   None of the other assessors can be satisfactorily explained by this preference dimension, but we can accommodate them by introducing a second dimension independent of the first.

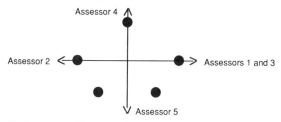

D)   Finally, assessor 6 bears no relation to any of the first five, but can still be accommodated by introducing a third dimension independent of the other two which would displace products A and E into the plane of the page, and B, C and D above it.

Figure 6.4 Simple demonstration of internal analysis.

order of variance explained, later dimensions are generally of minor importance, although they are always available for inspection.

Having generated a multidimensional preference map based on the acceptance data, it is possible to test the hypothesis that this derives from the sensory differences by examining the correlation of the preference dimensions with sensory, instrumental or other product data. In this respect the analysis is the exact opposite of external analysis. Note that direct regression against basic sensory data may not be sufficient to adequately explain the dimensionality, and that attribute interactions and quadratic effects may also need to be taken into account.

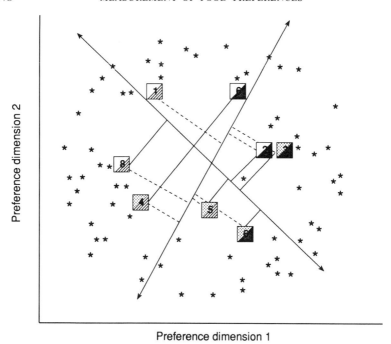

Preference dimension 1

Figure 6.5 Internal preference map obtained from consumer ratings of eight reformed steaks. Products 1, 2, 7 and 8 were high fat; products 1, 8, 4 and 5 were high salt. Asterisks show the directions of increasing preference of individual consumers. Projecting steak positions on to directions, as shown, gives estimate of individual consumer rankings that lie on those directions.

In certain circumstances, such as factorially designed experiments, it may be possible to forgo interpretation of the dimensionality of the preference space using objective data, since a knowledge of the product formulation linked with visual inspection of the maps may enable the experimenter to identify the desired formulation directly.

Such was the case in a study of restructured steaks where four factors (at two levels) were combined in eight products following a ¼ replicate of a 2^5 factorial design. The factors examined were fat content (12% versus 20%), salt content (0.5% versus 1.0%), tempering (long versus short) and blend time (12 mins versus 6 mins). The resulting internal preference map is shown in Figure 6.5. Visual inspection shows that two factors, fat and salt, are clearly associated with the first two preference dimensions, and are, therefore, of greatest importance in determining consumer acceptability of the restructured steaks. This figure also demonstrates another important feature of preference mapping. The product mean scores for overall liking revealed no significant differences between the eight products, and there were no significant differences across known socio-demographic sub-groups. On the basis of this conventional type of analysis one might have been led to believe that, to this sample of consumers, fat content and salt content were irrelevances. However, internal preference analysis demonstrated that this was not true. The majority of respondents were significantly fitted on dimensions 1 and 2 and do react to fat and salt. The problem with the data – though problem is perhaps not the right word for this situation – is that the consumers were highly segmented and equally distributed around the various sectors of the preference space. Thus, those in the top left sector preferred higher salt/higher fat, whereas those in the bottom right section preferred lower salt/lower fat, and the acceptance scores given by these groups neutralise each other in the aggregated data so that no net 'winners' or 'losers' are observed. Such variation in natural preferences is a good basis for establishing multiple brands, each variant addressing the sensory needs of a particular sector of the market-place. This is a very different conclusion to that resulting from the traditional analysis of the data in the aggregated form.

6.3.2   Case study using internal analysis

A study was commissioned by the Technical Department of a UK brewery to determine which characteristics of lager beer were important in determining product acceptance, and then to establish whether there was scope for improving product performance through a change in product formulation. The brewer, through previous experimentation, had established that he could produce five variants of his standard lager, all of which were significantly different from each other as determined by in-house discrimination tests. On our suggestion he also included in the study six competitive lagers of similar original gravity. This would have the benefit of demonstrating the variation across his own products relative to known market products. By increasing the number of products we should also be presenting consumers with greater variation to stimulate choice, and also increase the confidence of our analyses through increased degrees of freedom.

   Descriptive sensory analysis was conducted on the 12 lager beers by a panel of 11 consumers previously screened for their sensory ability. The panellists derived their own vocabulary for describing the products' sensory characteristics, using reference materials wherever possible to aid definition of the various characteristics. The agreed profile consisted of 38 descriptions, nine recording aroma characteristics and 29 describing taste, mouthfeel and aftertaste sensations. The assessments were conducted in triplicate with product presentation randomised over the panel and over replicates to reduce order biases. Two-way analysis of variance was performed on the data for each attribute to test the independence and presence of effects due to the different samples, the assessors, and also interactions between these two factors.

The consumers recruited for the acceptance test were male lager drinkers, aged 18 to 34 years, from C1 and C2 social groups, all of whom claimed to drink at least 6 pints of draught lager per week on licensed premises. The target sample size was 240, split equally over two regions of the UK, Northern England and Midlands, and it was also intended that half the sample would claim to be regular drinkers of the client's own brand.

Assessment of the lagers by consumers took place over two 90-minute visits to central location venues. Because of the time commitment and the necessity of running the test schedules under fairly tight control, respondents were pre-recruited for the test, the sample being increased to 300 to allow for about a 20% loss over the days of testing. At the test venue respondents were seated at individual tables and on the first visit briefed about the nature of the test being conducted (explanation of the procedure and the test questionnaires being used). Respondents were also reminded of the legal implications of driving motor vehicles after consuming alcoholic beverages.

Respondents received the lagers according to a unique ordering, specified for each individual, that placed each lager in each of the 12 positions of sampling every 12 respondents (MacFie *et al.*, 1989). Respondents received six test samples in each session. Each sample was presented and assessed monadically (without cross product comparisons). Respondents were given bread and water between samples in order to cleanse their palates.

In both the sensory and consumer assessments, the lagers were served at a temperature of 7/8°C in black half-pint glasses, thus eliminating many of the visual differences between the products.

Consumers were asked to rate the acceptability of the lagers on a 9-point category scale (dislike extremely–like extremely) and because of the need to relate the data to previous research, the client insisted on the use of 'Just About Right' (JAR) scales for the assessment of flavour strength, body, sweetness, bitterness, alcoholic strength, and carbonation. Internal preference mapping on the overall liking (acceptance) data was carried out using a GENSTAT program available at the Institute of Food Research.

6.3.2.1 Results. A total of 225 respondents completed the study assessing all 12 lager products. Analysis of variance revealed that the products were significantly different in terms of overall liking ($p = 0.001$) with a range of mean scores from 5.99 to 4.75, as shown in Table 6.1.

Similarly all the product attributes, with the exception of gassiness, revealed statistically significant differences between the products. Using

Table 6.1 Mean acceptance scores for the 12 lager products

Product	Acceptance score (scale 1 to 9)
Brand A	5.99
Variant 1	5.80
Variant 3	5.69
Variant 2	5.64
Brand B	5.63
Brand C	5.55
Standard Product	5.55
Brand D	5.32
Brand E	5.28
Brand F	5.16
Variant 4	5.06
Variant 5	4.75

Product scores within brackets not significantly different $(LSD_{95} = 0.40)$.

the JAR scales all the products were described as being too bitter in taste, and all of them were said to be lacking in flavour strength, body, sweetness and alcoholic strength. None of the attributes scores showed significant correlations with overall acceptance, though flavour strength, body, sweetness and alcoholic strength were positively correlated with acceptance, and bitterness negatively correlated.

Two-way analysis of variance conducted on the attribute scores relating products with a range of respondent factors including age, social class, region, brand use, weekly volume of lager and frequency of drinking, revealed many differences in the magnitude of scoring by the various respondent groups. However, no significant interaction between products and the various respondent factors was identified. Thus traditional analysis and examination of the aggregated data would have led us to conclude that all respondent sub-groups tended to view the products similarly.

Figures 6.6a and 6.6b illustrate the results of internal preference analysis, generated from the individual 'overall-like' scores given by the consumers. For each individual the end point of a linear vector, representing increasing acceptance, is shown. The first three preference dimensions were able to show a significant fit for 70% of the participating respondents.

The preference space defined by axes 1 and 2 shows the lagers split into 2 groups with brands A, B, D, and variants 2 and 5 positioned to the upper left sector of the space, and brands C, E, F and variants 3 and 4 positioned to the lower right sector. The overall direction of acceptance is towards the bottom left quadrant, therefore within each of the previously mentioned groups the lagers appear to be ranked along the overall acceptance dimension with brand A and variant 2 performing much better

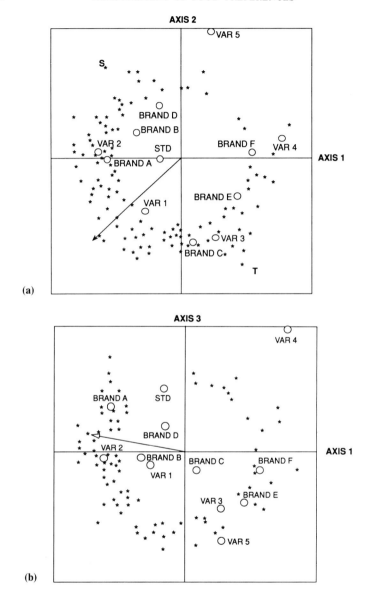

Figure 6.6 Demonstration of internal analysis. Product configuration generated from lager acceptance data. STD = current lager, vars 1 to 5 = experimental brews, brands A to F = competitive lagers, significant subject preference vectors shown as (∗), and direction of acceptance for total sample as →. (a) Preference dimensions 1 and 2. (b) Preference dimensions 1 and 3.

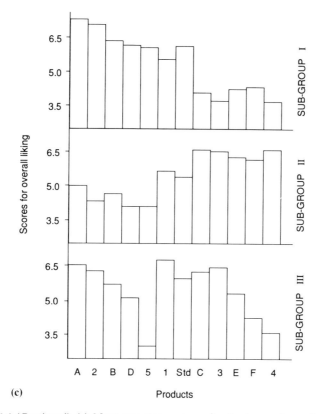

Figure 6.6 (Continued) (c) Mean acceptance scores for the lagers shown for each of the three preference sub-groups defined by similarity of preference in (a) and (b).

than variant 5 in the top group, and brand C and variant 3 performing much better than brand F and variant 4 in the bottom group. The standard product and variant 1 tended to fall between the two groups, with variant 1 further along the main acceptance dimension than standard lager. Figure 6.6b demonstrates the effect of the third preference axis, a dimension of preference that has little relationship with overall acceptability. The main effect of this dimension is to increase the separation between brand F and variant 4 which appeared to be close in Figure 6.6a, and obviously to further separate variants 4 and 5. Thus, by examining only the acceptance scores of the respondents, one is able to generate visible relationships or structure in the way that respondents perceive the products.

As in the previous figures describing the external analysis vector model, individual respondents are shown as an asterisk, each asterisk representing the end point of one respondent's acceptance vector. The direction of

Table 6.2 Acceptance scores for two opposing subjects

Product	Subject S	Subject T
Variant 5	9	3
Brand D	7	3
Brand A	8	5
Variant 1	8	5
Brand B	7	4
Variant 2	7	4
Standard lager	6	4
Brand F	5	5
Brand E	4	5
Variant 4	4	6
Variant 3	3	7
Brand C	2	8

the acceptance vector is visualised by drawing a line from the centre to an asterisk, the distance of any asterisk from the centre indicating how well that individual is fitted (i.e. how much variance is explained). Only those respondents who are significantly fitted (95% confidence) are shown on the plots.

The relationships between the individuals and the preference space can best be described by specific examples. Two individuals are shown in Figure 6.6a, they are labelled 'S' and 'T' and, because of their positions in the preference space, would appear to have opposing views about what they like. Their actual acceptance scores are shown in Table 6.2.

Subject S demonstrated a correlation of 0.885 with the first two preference dimensions, and his acceptance vector could account for 78% of the variance in his data. Subject T demonstrated a correlation of 0.860 and 74% of his variance explained by acceptance vector. Note that these vectors are not exact representations of each individual's scores, but are regressions onto the preference dimensions demonstrating a good fit of the original data (i.e. accounting for a significant proportion of variance). Both these subjects are fitted with greater than 99.9% confidence.

Clearly the majority of those respondents who can be fitted are negatively correlated with axes 1 and 2, tending to cluster into the lower left quadrant. However, the respondent sample is not homogeneous in its opinion of lager quality, with evidence of segmentation along all three axes. Thus, looking at axes 1 and 2 in Figure 6.6b, those respondents in the top left quadrant (referred to as sub-group I) are characterised by preferences for brands A, B, D and variants 2 and 5 and rejection of brands C, E, F and variants 3 and 4. Those respondents in the bottom right quadrant (referred to as sub-group III) have the opposite view of the products. Those respondents in the bottom left quadrant (sub-group II) represent the majority response of the sample (and reflect the overall result) in preferring variants 1, 2 and 3, and brands A and C, and rejecting variants 4 and 5.

Histograms of the product acceptance scores for the three major respondent sub-groups are shown in Figure 6.6c, emphasising their different views of the products. These respondents account for 45% of the total sample. A further 4% are fitted in the top right quadrant, clearly a minority view of product acceptance, whilst a further 20% are explained using preference axis 3, either alone or in conjunction with axis 1 or axis 2. The 30% of respondents, who are not statistically fitted on axes 1 to 3, show relatively little variation in their scores for lager acceptance.

Having segmented the respondents by similarity of preference, using the quadrants defined by the first two axes, one would like to ascertain whether it is possible to characterise the respondents within the subgroups, and perhaps gain an insight into whether their background or lifestyle might predict consumer preference. Table 6.3 summarises the demographic information recorded for the respondents at the time of recruitment. The most important observation to make is that no single demographic factor can be used to describe any particular sub-group. This is not too surprising, since two-way analysis of variance revealed no significant product × respondent interactions.

Table 6.3 Socio-demographic details of respondents within preference sub-groups

Respondent demographic factors	Total sample (%)	Preference sub-groups (%)			
		Sub-group I	Sub-group II	Sub-group III	Non-fitted
Social class					
C1	26	27	31	29	21
C2	74	73	69	71	79
Age					
18–24 years	59	67	51	58	59
25–34 years	41	33	67	42	41
Region					
North	44	47	52	29	43
Midlands	56	53	48	71	57
Frequency lager consumption					
4 or more times/week	50	58	40	46	59
2 to 3 times/week	50	42	60	54	41
Weekly volume consumed					
20 or more pints	39	45	33	46	41
5–19 pints	61	55	67	54	59
Other drinks consumed					
Bitter	15	0	16	33	14
Stout	4	0	4	25	1
Wine	16	6	18	21	17
Whisky	24	21	18	43	21

However, there are high or low frequency biases in distribution associated with certain sub-groups, but each factor type is generally found within each preference sub-group. Thus sub-group I has a bias towards 18- to 24-year-old lager drinkers, who are also slightly more likely to drink lager on four or more occasions each week. By contrast, group II has slightly more 25- to 34-year-old drinkers, who are more likely to be drinking lager on fewer occasions and, consequently, consume a smaller volume of lager per week. Group III has a distinct Midlands bias, but otherwise is fairly typical of the overall sample.

Interestingly, sub-group III contained most bitter and stout drinkers likely to be drinking wine and whisky. By contrast, sub-group I respondents made fewer claims regarding consumption of other alcoholic beverages, and none of them claimed to be drinking bitter beers or stout beers. Lists of brands of lager consumed within the last month or consumed most often also failed to reveal any natural segmentation of the respondents corresponding to their preference behaviour. Thus regular drinkers of the standard lager were fairly equally distributed across the preference sub-groups, and there was no indication of greater frequencies of brands C, E or F drinkers in sub-group III, nor of brand A, B or D drinkers in

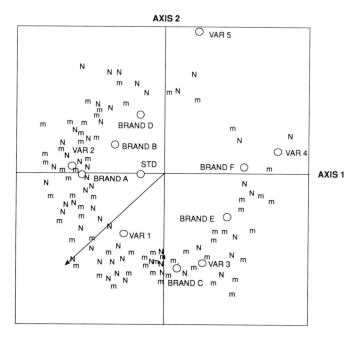

Figure 6.7 Internal analysis on lager acceptance data. Individual subjects who are significantly fitted on preference dimensions 1 and 2, labelled by region (N = North of England; M = Midlands).

sub-group I. An alternative approach to examining the demographic associations within the preference sub-groups is to attach appropriate labels to respondents as shown in Figure 6.7. Although a useful tool for highlighting any interesting features of respondent segmentation, it tends not to be quite so useful in identifying important differences, since the differences in distribution across the preference space need to be quite marked in order to be visually distinct.

In most aspects the 30% of respondents not fitted by the preference mapping are identical to the overall sample, both in terms of demographic factors and products consumed. As described in an earlier section, there are three possible reasons for lack of fit by the model. Firstly, they could quite genuinely have felt that the lagers were equally acceptable, perhaps as a result of poor sensory acuity; secondly, they may have judged acceptance on criteria that the majority of respondents do not use and, therefore, are ignored as a collection of many minority groups; or finally, they may not have understood the test or how to use the scales, and simply generated random scores.

Having generated a preference map, using internal analysis, observed the relationship between products and segmented respondents by similarity of acceptance/preference, in order to make full use of this information it is now essential that the preference dimensions are interpreted. This is done by identifying those product characteristics which are related to the preference map, and testing whether they may have been used by respondents in making their choices on product acceptability. If there are several characteristics associated with a preference dimension, one should not automatically assume that they are all important in explaining acceptance. Obviously, they will have some association with acceptance, but this association may be entirely coincidental. However, it is true that characteristics which are not related to the preference space are unlikely to play a major role in determining consumer acceptance.

Figure 6.8 illustrates how the consumer scoring of lager attributes are related to the preference space. The directions of the attribute vector lines are obtained by regressing the attribute mean scores onto the product coordinates for those preference axes under consideration. Attributes that are close together indicate positive correlations, those at right angles to one another are uncorrelated, whilst those in opposite directions indicate negative correlations. The consumer data, as shown in this figure, are fairly unidimensional. The more acceptable products tend to have been described as sweeter, more gassy, more alcoholic, with more flavour strength and body. The less acceptable products have been described as more bitter. Although attributes on face value give a good explanation of the overall acceptance scores, the reasons why a two-dimensional preference map was required to explain the dimensionality of the preference space are not brought out.

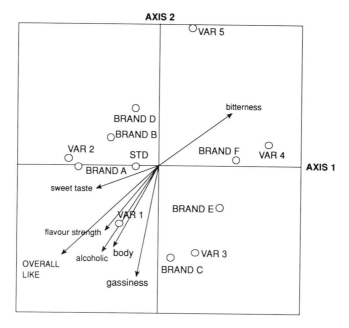

Figure 6.8 Interpretation of the lager preference dimensions (axes 1 and 2) using the consumer ratings of product attributes.

A better way of interpreting the preference space is to use more objective sensory data generated by a panel trained to quantify their perceptions without the biases introduced by likes and dislikes. This information is regressed on to the product coordinates on each preference axis, as shown in Figure 6.9. Characteristics which are positively associated with lager acceptance are fruitiness (citrus fruits and fermenting fruits/over-ripe), fragrant and perceived alcohol, and to a lesser extent body and sweetness, whereas vegetable characteristics and graininess (both sulphurous in origin), soapiness and, to a lesser extent, bitterness are negative characteristics. Actual correlations between the mean acceptance scores and the sensory attribute scores are shown in Table 6.4.

However, this is only part of the interpretation, since individual respondents did not perform identically and showed considerable variation on secondary axes 2 and 3. Thus respondents with preference vectors clustered into the top left quadrant (sub-group I) would appear to be less affected by differences in fruitiness, and instead are looking for beers with more malt character, slightly more toffee flavour, more sweetness and body. They would also appear to be more tolerant of vegetable-like/cabbagy qualities. They seem to be critical of the more bitter beers and their associated hoppiness. Respondents in sub-group III (lower right

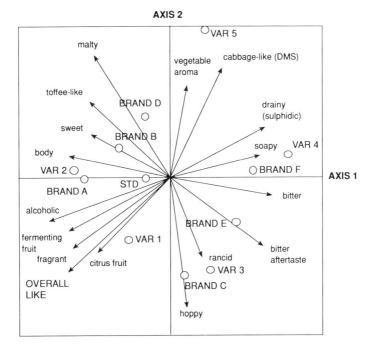

Figure 6.9 Interpretation of the preference space defined by axes 1 and 2, using sensory data generated by a trained panel.

Table 6.4 Correlations between lager sensory characteristics and mean consumer acceptance

Sensory attribute	Correlation coeff.	Confidence level
Alcoholic	+0.81	<0.01
Drainy	−0.77	<0.01
Fermenting fruit	+0.74	<0.01
Fruity citrus	+0.70	<0.05
Fragrant aroma	+0.69	<0.05
Soapy	−0.67	<0.05
Cabbage-like	−0.65	<0.05
Body	+0.53	<0.1
Bitterness	−0.49	NS

quadrant) could be the exact opposite, exhibiting a greater preference for the more bitter, less sweet beers and, perhaps, critical of malty, vegetable-like qualities of lagers. The preferences of these groups relate well with the observations concerning other alcoholic beverages consumed. Sub-group III respondents showed a much greater tendency to consume the more bitter 'bitters' and 'stout' products and to consume more varied alcoholic beverages, whereas none of the respondents in sub-group I claimed to consume bitters and stouts and might, therefore, be expected to prefer less bitter beers generally.

On the basis of these relationships it is possible to suggest why the lagers perform the way that they do in a blind acceptance test, and to indicate how the product performance can be further improved. In this study, overall acceptance seemed to be positively influenced by fruitiness which seemed to be associated in some way with perceived alcohol content. Negative associations were indicated between acceptance and the sulphury nature of the lager (vegetable, drainy qualities) and soapiness, both of which are important descriptors associated with 'traditional' lager flavour (Clapperton and Piggott, 1979). The results suggest that the brewer should increase the fruitiness of his lager or decrease the sulphur character, or both. Note that, because of the nature of this analysis, it is not possible to differentiate between the effects of fruitiness and sulphury flavour, nor to be more specific than to indicate the direction of the change. Further study would be necessary to establish the importance of the flavours and to determine ideal levels of both or either flavour constituents. The brewer also has the potential of aiming his product at one or more market sectors by modification of the bitterness of his lager. A slightly less bitter lager would have greater appeal to the younger lager drinker, whereas a slightly more bitter lager would have greater appeal to the more 'experienced' drinker with a wider repertoire of beverages consumed.

6.4 Advantages and limitations of preference mapping

Having described the alternative approaches to preference mapping and illustrated their use through case studies, it should be apparent that both internal and external methods of analysis offer the same advantages to the researcher.

First of all, they give a better insight into the underlying reasons behind the consumers' decision-making when assigning acceptability scores to products. The logic behind this is that, if consumers can differentiate between products in terms of acceptance (i.e. they like some products more than others), then respondents must have perceived some characteristic(s) present in the various products and made their judgement based on the relative intensities of these characteristics, even though they may not be able to describe their sensations to the researcher. If we know how the products differ in terms of their sensory characteristics (or other objective measurements), then one can determine which of the various differences are important in determining consumer acceptance.

Secondly, they allow the researcher to view the acceptability response at the individual level and, thus, see how individuals' opinions vary or cluster into similar groups. This is a much more revealing presentation of acceptability data than simply looking at mean scores. For example, two groups of consumers could have opposite views about the products under test, such that, when combined, would generate similar mean scores for all products. Although this would suggest that there was no difference in acceptability between the products, this would clearly not be true. Preference mapping, by showing each individual response, would indicate the different opinions of the two groups very clearly.

However, the two techniques are not totally without drawbacks. The main problem or worry associated with both types of analysis is that the interpretation is entirely dependent upon the quality of the external data. It is unwise to ignore product variation that exists, but which may not seem to relate to the intended experimental design, since one can be sure that some consumers will react to that variation, no matter how unintentional its presence. Incomplete sensory data increase the risk of an incomplete explanation of the consumer response. Therefore one must have an accurate record of the perceptual variation that exists across the products, particularly with respect to products as presented to the consumers. This usually means that the products should be of similar age, ideally taken from the same stock, and served in identical conditions.

Another problem specifically associated with external analysis is that one usually reduces the external data to a fewer number of dimensions using techniques such as principal component analysis. Therefore the external data used in the preference analysis are usually the product coordinates on the first two or three components.

However, this assumes that the greatest product variation is likely to be more important to the consumer than minor variation, and this may not always be the case. In a previous publication describing preference mapping of fragrances from a household product (Nute et al., 1989) we demonstrated that the coordinates on components 2 and 3 gave a better explanation of consumer preference than any configuration involving component 1. Had the important factor been component 4 or 5, this would have been missed by our usual practice of including the first three. Although one can avoid this problem by including the product coordinates for all relevant sensory dimensions (given that we have a workable definition of relevant dimensions), this greatly increases the number of component combinations that need to be examined. Taking all Prefmap models (phases IV, III, II, I) into account escalates the task of interpreting consumer response dramatically when more than three components are input.

Internal analysis does not suffer from this drawback because it operates directly on the preference data and automatically highlights the most important dimensions of preference. In practice three dimensions are

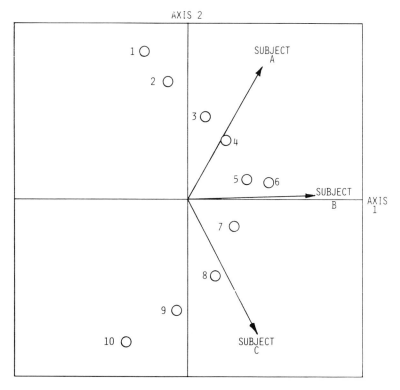

Figure 6.10 Product configuration from internal preference analysis indicative of ideal point interpretation (products labelled 1 to 10).

usually sufficient to describe the acceptance behaviour of a large majority of respondents; however, other preference dimensions are always available for inspection as part of the standard analysis.

One criticism that is sometimes directed towards internal analysis is that the dimensions are generated on the basis of individual *linear* preference vectors, that there is no scope for demonstrating ideal points. In the main this is true, although there are product configurations resulting from internal analysis which are indicative of quadratic relationships and an ideal point. Figure 6.10 illustrates this. The product configuration will typically look 'V-shaped' or 'U-shaped' and will have relationships with external data which are at right angles to the direction of the V. In the example shown one might find that sweetness is well correlated with axis 2, such that product 1 is most sweet, and product 10 is least sweet. Respondents in the upper right quadrant, such as subject A, would seem to prefer sweeter products, giving high scores for products 1, 2, 3 and 4, and rapidly decreasing scores for products 7, 8, 9 and 10 in which sweetness is in decline. Subjects in the lower right quadrant would seem to prefer less sweet products (i.e. subject C). However, those respondents positioned close to axis 1, such as subject B, find products 5 and 6 most acceptable and are increasingly critical of products containing greater or lesser intensities of sweetness. For these respondents it is possible to establish an ideal point for sweetness.

Another drawback which relates to the practical aspects of preference mapping is that each subject has to assess all of the samples. Missing value routines for eigen-vector analyses are available (e.g. Beale and Little, 1975), but are not widely implemented. Currently subjects who have missing values are removed from the data prior to the analysis; and the practical tip that avoids unnecessary waste of resources is to check respondent questionnaires prior to removing a sample and replacing it with another. For external analysis missing values are not so critical, since the scores of each subject are fitted separately, but direct comparability across respondent preference segments may be lost.

Another problem with preference mapping techniques, which seems to have been poorly addressed, is how one can assess the significance of the models derived. How reproducible and robust are the maps that are generated? Some researchers approach the problem by using an additional sample of between 20–40% of subjects who assess half the products twice and thus provide an estimate of the variance for the whole sample. However, there are considerable cost implications in doing this.

Another approach, which is certainly less costly, is to duplicate some of the samples and treat them as additional test products. This is particularly useful in internal analysis since one should always see the replicated samples close together if the test is conducted correctly and the subjects are performing well. The greater the number of replications, the greater the confidence in the resulting configuration.

In a recent development Takane and Shibayama (1991) use the Boot-strap method of Efron (1979) to derive 95% confidence regions for the stimulus points found by internal preference mapping. In an interesting extension, these authors also cater for the case where there is external information on both subjects and stimuli.

6.5 Aspects of conduct

A highly important requirement for preference mapping is that order effects are minimised so that they have little impact on the overall consumer response. In our experience simple randomisation of order is inadequate. We have adapted an algorithm given by Williams (1949), which produces designs balanced for the effect of a single preceding treatment every n replications if n is even, and every $2n$ replications if n is odd (MacFie et al., 1989).

Table 6.5 Eight treatments: 48 replicates in six balanced blocks of eight consumers

Consumers	Order of assessment							
	1	2	3	4	5	6	7	8
1	5	4	8	7	1	2	6	3
2	8	5	1	4	6	7	3	2
3	4	7	5	2	8	3	1	6
4	3	6	2	1	7	8	4	5
5	7	2	4	3	5	6	8	1
6	6	1	3	8	2	5	7	4
7	2	3	7	6	4	1	5	8
8	1	8	6	5	3	4	2	7
9	1	7	5	8	3	4	6	2
10	7	8	1	4	5	2	3	6
11	6	3	2	5	4	1	8	7
12	8	4	7	2	1	6	5	3
13	5	1	3	7	6	8	2	4
14	2	6	4	3	8	5	7	1
15	4	2	8	6	7	3	1	5
16	3	5	6	1	2	7	4	8
17	2	6	5	1	7	3	4	8
18	7	5	4	2	8	6	3	1
19	5	2	7	6	4	1	8	3
20	1	3	6	8	2	4	5	7
21	6	1	2	3	5	8	7	4
22	4	7	8	5	3	2	1	6
23	8	4	3	7	1	5	6	2
24	3	8	1	4	6	7	2	5
25	6	1	3	4	8	5	7	2
26	8	3	7	6	2	1	5	4
27	5	2	4	7	1	8	6	3
28	2	7	5	8	4	3	1	6
29	1	4	6	5	3	2	8	7
30	3	6	8	1	7	4	2	5
31	7	8	2	3	5	6	4	1
32	4	5	1	2	6	7	3	8
33	4	1	2	7	5	6	8	3
34	7	6	1	3	4	8	2	5
35	3	8	6	5	7	2	1	4
36	1	7	4	6	2	3	5	8
37	5	2	8	4	3	1	6	7
38	6	3	7	8	1	5	4	2
39	2	4	5	1	8	7	3	6
40	8	5	3	2	6	4	7	1
41	6	1	5	8	3	4	2	7
42	3	5	2	6	7	1	4	8
43	5	6	3	1	2	8	7	4
44	1	8	6	4	5	7	3	2
45	8	4	1	7	6	2	5	3
46	7	2	4	3	8	5	1	6
47	4	7	8	2	1	3	6	5
48	2	3	7	5	4	6	8	1

Table 6.5 gives an example for eight treatments in six balanced blocks of eight consumers. To test the design, the reader is invited to confirm that, in every block, each sample occurs once in each position and each pair of samples occurs twice, once in each order (i.e. 5, 4 and 4, 5).

Order effects may also become apparent if a large test design requires subjects to assess products over more than one session. This scenario is fairly frequent since, even with the most innocuous of products, there is a limit to the number of samples that one can expect a consumer to assess without sensory fatigue or other performance degrading factors coming into play. With products in which physiological/psychological changes may occur, for example through ingestion of alcohol, then multi-session testing is almost compulsory. One way of reducing the impact of order effects, particularly the common difference observed between first and subsequent products, is to introduce a dummy sample in the first position. The scores of this sample are not input to the preference analysis.

However, in many tests it may be possible to conduct all the product evaluations in one session, especially since one only requires subjects to rate each product once (i.e. how much do you like the product overall?). Rating of product acceptance, and perhaps collection of spontaneous likes and dislikes (to stop subjects who wish to amplify their feelings from becoming frustrated) on say 12 products, is a fairly undemanding task provided subjects are not rushed. It is certainly much easier than rating one or two products on say 15 to 20 product attributes. We would generally feel happier with just one question, allowing subjects to react instinctively towards each product rather than analytically. We then use unbiased external data to provide the analytical interpretation.

References

Beale, E.M.L. and Little, R.J.A. (1975) Missing Values in multivariate analysis. *J.R. Statist. Soc. B*, **37**, 129–45.

Bockenholt, I. and Gaul, W. (1986) Analysis of choice behaviour via probabilistic ideal point and vector models. *Applied Stochastic models and Data Analysis*, **2**, 209–26.

Carroll, J.D. (1972) Individual differences and multidimensional scaling, in R.N. Shepard, A.K. Romney, and S.B. Nerlove (eds) *Multidimensional Scaling: Theory and Applications in the Behavioral Sciences*, Vol. 1, Seminar Press, New York, pp. 105–55.

Chang, J.J. and Carroll, J.D. (1968) How to use MDPREF, a computer program for multidimensional analysis of preference data. Unpublished report, Bell Telephone Laboratories.

Clapperton, J.F. and Piggott, J.R. (1979) Differentiation of ale and lager flavours by principal components analysis of flavour characterisation data *J. Institute of Brewing*, **85**, 271–4.

De Soete, G. and Carroll, J.D. (1983) A Maximum Likelihood Method for fitting the wandering vector model. *Psychometrika*, **48**, 553–6.

Efron, B. (1979) Bootstrap methods: another look at the jack-knife. *Annals of Statistics*, **7**, 1–26.

Girard, R.A. and Cliff, N. (1976) A Monte-Carlo evaluation of interactive multidimensional scaling. *Psychometrika*, **44**, 69–74.

Jedidi, K. and DeSarbo W.S. (1991) A stochastic multidimensional scaling procedure for the spatial representation of three-mode three-way pick any/J data. *Psychometrika*, **56**, 471–94.

Jones, P.N., MacFie, H.J.H., and Beilken, S.L. (1989) Use of preference mapping to relate consumer preference to the sensory properties of a processed meat product (tinned cat food). *J. Sci. Food Agric*, **47**, 113–23.

Kakamura, W.A. and Srivastava, R.K. (1986) An ideal-point probabilistic choice model for heterogeneous preferences. *Marketing Sciences*, **5**, 199–218.

King, B. and Arents, P. (1991) A statistical test of consensus derived from generalised procrustes analysis of sensory data. *J. Sensory Studies*, **6**, 1, 37–48.

MacFie, H.J.H., Bratchell, N., and Greenhoff, K.G., and Vallis, L.V. (1989) Designs to balance the effect of order of presentation and first-order carry-over effects in hall tests. *J. Sensory Studies*, **4**, 129–48.

Nute, G.R., MacFie, H.J.H., and Greenhoff, K. (1989) Practical application of preference mapping, in D.M.H. Thomson (ed.) *Food Acceptability*, Elsevier, London, pp. 377–86.

Piggott, J.R. and Sharman, K. (1986) Methods to aid interpretation of multidimensional data, in J.R. Piggott (ed.) *Statistical Procedures in Food Research*, Elsevier, London, pp. 181–232.

Schiffman, S.S., Reynolds, M.L., and Young, F.W. (1981) *Introduction to Multidimensional Scaling*, Academic Press, New York.

Takane, Y. and Shibayama, T. (1991) Principal Component Analysis with external information on both subjects and variables. *Psychometrika*, **56**, 97–120.

Whelehan, O.P., MacFie, H.J.H., and Baust, N.G. (1987) Use of individual differences scaling for sensory studies: Simulated recovery of structure under various missing value rated and error levels. *J. Sensory Studies*, **1**, 1–8.

Williams, E.J. (1949) Experimental designs for the estimation of residual effects of treatments. *Austral. J. Sci. Res. Ser. A*, **2**, 149–68.

7 An individualised psychological approach to measuring influences on consumer preferences
M.T. CONNER

7.1 Introduction

A common assumption in the sensory analysis of foods and drinks is that the assessment of acceptability must be separated from discriminative and descriptive testing. This has led to the standard requirement that sensory analysis avoids preference methods and uses trained panellists (Sensory Evaluation Division of the Institute of Food Technologists, 1981). However, research on the psychology of food acceptability indicates that this need not be so (Booth *et al.*, 1983; Booth and Conner, 1990). The contribution of any factor to the acceptability of a food or drink can be measured in terms of a consumer's discriminative performance in rating acceptability, provided the test situation is designed to avoid various biases. This chapter describes the theoretical model underlying such an approach, the methodological principles guiding measurement, how such acceptability ratings can be converted to measures of discriminative performance, and how these measures of individual performance can be combined across consumers to give meaningful estimates of market response to product formulation changes. Finally, extensions to the model are described which take its area of application beyond the single-factor sensory determinant of food acceptance from which it was initially developed.

7.2 Measuring individual consumer preferences

The sales and intakes of foods and beverages are the result of the choice behaviour of individual consumers who buy and consume particular products. It is the measurement of the strength of influence of those factors determining choice that this chapter is concerned with. In order to understand the influences on sales and intakes of foods and beverages we must estimate the parameters of the factors influencing choice within the individual consumer. These influencing factors can be broadly split into three: sensory, attitudinal and physiological. Sensory factors include the taste, smell and texture of the product and are based on the physicochemical characteristics inherent to the food itself. Attitudinal factors include the effects of packaging and marketing as well as broader influences such as social factors (e.g. health beliefs) on the acceptability of the product. Physiological factors include consequences of consuming the food such as stomach distension which affect feelings of fullness. Each of these influences will interact with one another. So, for instance, sensory influences may be tempered by advertising about the physiological effects of the product, e.g. its 'slimming' qualities. Thus, while a breakfast cereal may be less sweet than the consumer might ideally choose, this lack of sweetness may be offset by the healthy 'slimming' advertising of the product, setting up an expectation of a less-sweet cereal. These interactions imply that an individual's food choice cannot be fully understood simply by examining discrete influences upon choice (section 7.5).

The traditional way of measuring influences upon acceptability is a consumer popularity poll amongst variants of the product which differ in the factors of interest (e.g. sweetness, packaging, etc.). The procedures usually involve some form of preference ranking or rating: consumers may be asked to choose the most preferred of pairs of samples or to rank the variants in order of preference. Of the rating techniques, the nine hedonic categories of like and dislike (Peryam and Girardot, 1957) are the most widely used (Meiselman, 1984), while the use of 'magnitude estimation' techniques (scoring under ratio instructions) for acceptability has been advocated by some (see Moskowitz, 1982, 1983 for details). These rating techniques could also be used for measurement of the influences upon acceptance within individuals. However, they are usually not. The results are usually simply combined across the panel of consumers tested. This approach can give only the relative popularities of the tested variants or perhaps a product-attribute preference space extracted from the panel's data as a whole. As they do not consider how the varied attribute might determine acceptability within each individual they cannot relate attribute level to acceptability in any causal way, no matter how complex the statistical analysis employed (Booth, 1988).

What we need to do in order to measure these causal influences upon acceptance within an individual is to present likely determining factors varied by known amounts (within ranges that are acceptable to each individual consumer) and to observe the effects on the individual consumer's choice of product or brand. The differences in choices, caused by the differences in varied factors, are psychophysical functions. The analysis of such data differs radically from current practice in both sensory evaluation and consumer preference research. Instead of statistically analysing the grouped data for differences and patterns, this approach tests for causal relationships within individual consumers' behaviour. The measurement of the influences upon acceptance within the individual is guided by two interrelated theories. The first describes how food preferences are acquired, while the second concerns the relationship between acceptability judgements and levels of the determining factors.

7.2.1 *Acquisition of food preferences*

The conjunctions of particular levels of the factors which influence food choice is learned from early life onwards and familiar levels come to be preferred. Any deviation from this learned complex is likely to be less preferred, whether there is an excess or a deficit of any one particular characteristic (Booth *et al.*, 1987). This will be equally the case for characteristics that are perceived as major attributes of foods (e.g. sweetness in confectionery) as for characteristics that are little noticed (e.g. sweetness in soups). The only difference in attributes will be in the size of the effect that a minor deficit or excess of the characteristic has on overall preference for the food or drink, with differences in major attributes more likely to produce larger preference changes.

The importance of mere exposure to objects in determining preferences for such objects has been studied for some time in social psychology (e.g. Bornstein, 1989). The way in which factors influence choice of foods and drinks in familiar situations is similarly learned via exposure by a variety of mechanisms, including familiarisation, association with social and nutritional benefits and innate sensory reinforcement (Booth, 1985, 1990; Booth *et al.*, 1976, 1982; submitted b). Hence, palatability will adapt to the specific foods or drinks to which the individual is frequently exposed. Palatability of a particular food item has indeed been found to be directly related to the frequency of its recent usage for baby food in infants (Harris and Booth, 1987), novel cheeses and fruit in young children (Birch and Marlin, 1982) and unfamiliar fruit drinks in adults (Pliner, 1982). The strongest evidence, however, comes from the work of Mary Bertino and colleagues and concerns adaptation to differing levels of salt in the diet (Bertino *et al.*, 1982, 1986). In one study (Bertino *et al.*, 1982), a small group of individuals were encouraged to make a large decrease in their overall intake of salt (50%) and their changes in preference for differing levels of salt in soup and crackers was measured over time against a control group. The results showed how, during the course of several months, the experimental group gradually came to prefer lower levels of salt in these two foods compared to the control group who did not change.

The above theory of acquired food preferences specifies why individuals have most preferred levels of each of the factors relevant to choice and why deviations from these levels are less preferred. As such it provides the basis for a procedure which relates the level of a varied factor (intensity) in a particular food to its acceptability within an individual, in a fashion that actually measures the strength of influence of that factor on acceptance. On this theory, any individual faced with a food or drink will have a personally familiar formulation that they like most. Differences from the most preferred level in any readily perceived attribute are liable to reduce acceptability. The stronger the influence of that constituent on that person's attraction to that type of food, the more likely is a barely detectable deficit or excess to have an effect on selection amongst variants. The precise nature of the relationship of differences from most acceptable level to acceptability ratings is described by a psychophysical theory which has been named the acceptance triangle (Booth and Conner, 1990).

7.2.2 Relating determinants to food preferences: the acceptance triangle

Consumers mostly know and can say what they like and dislike, even when they find it difficult to say why. There are some characteristics of products for which the less there is the better (e.g. nasty off-flavour) or more there is the better (e.g. healthy) for the individual consumer. In these monotonic cases, linear regression can be used directly to identify the psychophysical acceptance function. However, for many factors, there is a most preferred level for any particular consumer in a given context. Products with more or less of this factor will be less acceptable than the product with the ideal level. The plot of acceptance responses against factor levels will thus be peaked. As the data are traditionally collected and plotted, this 'hedonic' function is an asymmetrical inverted U. However, the theoretical function is symmetrical and pointed in an inverted V (an isosceles triangle) if the units of the acceptance-determining factor are chosen to be equally discriminable and the consumer test has been correctly designed to avoid biases (Booth and Conner, 1990).

This inverted V function has been called an acceptance triangle (Figure 7.1). The triangle unfolds to give a straight line when the excess limb is reflected in the ideal-response horizontal (Figure 7.1). The characteristics of this unfolded function represent the strength of the mechanism (within the mind of the individual) by which the factor influences acceptance.

The acceptance triangle is a graphical representation of the psychophysical theory that decreases in acceptability from most preferred level are directly proportional to the number of discriminable differences from most preferred level of the determining factor, i.e. that judgments (ratings) are linear to discriminable differences. This is a version of Fechner's principle (Fechner, 1860) that expression of subjective magnitudes relies on the same process as discrimination of stimulus values.

The acceptance triangle represents a mental mechanism relating acceptability ratings to determinants of acceptance within an individual consumer. This is a quantitative scientific theory of behaviour, stating that decreases in acceptability from its maximum are directly proportional to the number of discriminable differences from ideal. Hence, unbiased

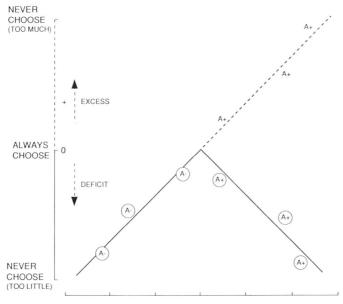

Figure 7.1 The acceptance triangle: indicating the predicted psychophysical relationship between measures of a varied factor scaled in equally discriminable units and an individual's acceptance ratings. Acceptance ratings (A− and A+) are linear against the number of discriminal differences from most preferred level, i.e. an inverted V-shaped function. The dashed line indicates how the inverted V can be unfolded into a straight line by reflecting the above ideal ratings (A+) in the always choose horizontal.

acceptability ratings should plot against the measure of the varied factor scaled in equally discriminable differences to give an inverted V, with equal slope(s) above and below maximum acceptability (Figure 7.1). The acceptance triangle (Figure 7.1, solid line) 'unfolds' into a straight line, because it has the same slope value above and below the ideal value of the determining factor (Figure 7.1, dashed line).

Overall acceptability may be determined by several discrete (independent) causes within the individual. Each such influence is represented by a separate acceptance triangle (section 7.5.1) which quantifies the effect of each discrete determinant on choice behaviour. Alternatively, a unitary factor could prove to be complex at another level of analysis. That is, the effects of variations in several determinants are sometimes merged into a single determinant of acceptability, such as a complex sensation or a generic attitude. In this case a single acceptance triangle could represent the effects of several interacting determinants.

To obtain such a measurement of causal strength, the appropriate scaling of the varied influence on acceptability will be those units which give equally discriminable differences in the factor. For aqueous solutions of sucrose in the ranges commonly used in foods, ratios of concentration units give equally discriminable differences (Schutz and Pilgrim, 1957). In such a case, acceptability ratings will plot linearly against a logarithmic scale of concentration. For other constituents of foods and drinks it may be necessary first to determine those units which give equally discriminable differences. The correct units will be those giving the tightest linear relationship between acceptability ratings and levels of the factor influencing acceptability. It is worth emphasizing that this linear relationship between acceptability and its determinants is a measure of the mental mechanism mediating this relationship. On this view, a failure to find a tight linear relationship indicates a failure to find the real basis underlying acceptability for that determinant in that individual. It is not a test of a type of theoretical modelling: it is a test of the investigator's understanding of that aspect of consumer behaviour.

According to this theory, the often observed inverted U or even asymptotic curve (Moskowitz et al., 1974) obtained from hedonic ratings of sweet and other stimuli are artefacts of a poorly designed test situation or scoring procedure or of analysis based upon premature aggregation of each individual assessor's data. To describe such indeterminate curves, optimisation procedures have to resort to polynomial regression (Moskowitz, 1983) and even then the distortion will remain. When such data are appropriately collected and analysed, on the other hand, the strength of an influence on acceptability lies in the characteristics of a mathematically determinate inverted V (Booth et al., 1987).

The danger from averaging the ratings by different individuals is that the grouped maximum response will be unrepresentative of the majority or perhaps even of anyone in the group (Pangborn, 1981). Even more to the point, there is no information on how many or how few individuals find an item with any particular level of the varied factor to be acceptable.

Also, averaging or forcing a consensus onto panel data will obscure any qualitative differences between individuals. This is particularly serious for acceptance functions which are the result of several factors interacting – because the interaction may be in unsuperimposably different patterns in different people. The consensus space or the distribution of average responses can quite misrepresent individuals who have other sorts of interactive patterns at peak acceptance.

Most fundamentally though, the panel mean or consensus cannot provide a measure of the strength of the causal processes actually operative in consumer behaviour. It is a person who thinks and feels, not a panel's popularity poll or the total sale in the market. Hence acceptance responses to known situations must be measured and analysed first individual by individual. However diverse the interactions producing them, the personal response spaces from a representative panel can then be summated to estimate the accumulated responses from a population.

7.2.3   Principles of consumer preference measurement

The above theory specifies how acceptability responses can be causally related to the factors determining acceptability. This section describes the principles guiding the collection of such acceptability data in order to avoid biasing effects which might distort the triangular relationship between acceptability and its determinants.

The basic method is to present likely determining factors varied by known amounts within ranges that are acceptable to each individual consumer and to observe the effects on the consumer's disposition to choose the product or brand. The differences in choices caused by the differences in conditions are psychophysical functions of the effects of the determining factors on individual acceptance. Combinations of psychophysical acceptance functions yield reliable and valid measures of interacting determinants of acceptance within each individual tested.

In order to obtain an undistorted psychophysical function (acceptance triangle) for any one influence on choice, an appropriate test design must be used to avoid biases. Two sorts of psychophysical principle should be applied.

First, the relevance and precision of the responses depend on the test conditions being sufficiently close to real use of the product to be familiar to the assessor. In any approach, sound predictions of consumer choice require the mimicking of the real-life situation. We must study real brands or foods in normal situations of purchase or use and record actual behaviour, or at least the ratings should be in relation to preference behaviour (e.g. always buy, never choose) and not in terms of subjective pleasures and quantities (e.g. like extremely). People are rather good at accurately reporting what they would do in a particular situation, while there is no reason to believe that imprecise expressions of private pleasure will be well related to actual preference behaviour. That is, the words on which the verbal assessment is anchored should refer to objectively agreed events (behavioural responses and familiar test item formulations). This is the principal reason for rejecting the widely used categories of like and dislike, or pleasantness or preference.

The acceptability ratings can be collected in either of two ways: as folded (uncharacterised) acceptability ratings or as unfolded (characterised) relative-to-ideal ratings. Folded acceptability ratings simply require an assessor to rate the acceptability of the product by marking a point between pairs of categories of overall acceptability, such as 'Always choose' and 'Never choose' or 'Always buy' and 'Never buy'. The only requirement is that pairs of descriptors are easily usable by the assessor for placing a response. These ratings can be converted from a peaked function to a straight line by reflecting the above-maximum acceptability responses in the most preferred axis (Figure 7.1). The concentration at which this unfolding is performed can be determined by iteration using the best linear fit of the data as criterion (Conner, 1989). Alternatively, if the assessor can say what is influencing her, she can be asked to assign samples to differences from the most acceptable level, stating whether the level of the perceived relevant factor is excessive or deficient (Booth et al., 1983). The overall acceptability ratings categorised as arising from excess are then scored as above ideal, providing an unfolded function.

Unfolded acceptability data are collected through the use of so-called ideal-relative ratings. These are particularly useful where one or a few easily recognisable characteristics are varying in the product. So, for instance, for ratings of sweetness in a beverage an assessor might rate presented samples between three ordered rating anchors, 'So little sweetness I'd never choose it', 'At this sweetness I'd always choose it', and 'So much sweetness I'd never choose it'.

The functions resulting from these two rating procedures have been found to be very similar. However, the first procedure avoids unnecessarily focusing the assessor's attention upon the varied characteristic(s), risking an overestimate of its (their) effects upon acceptance.

The second principle for precise measurement of the determinants of acceptance is that known sources of bias in the judgement are minimised. Biases in individual responses would tend to distort or displace the underlying psychophysical function so that it is not V-shaped. It is crucial that these various sources of bias are avoided in rating the acceptability of samples of a food or drink. Biases in such ratings have been demonstrated by experimental psychologists (e.g. Poulton, 1979, 1987, 1989) and procedures for minimising biases have been developed and applied in food research (Riskey et al., 1979; Booth et al., 1983; McBride, 1985; Conner et al., 1987, submitted).

For example, the bunching of responses in any part of the response dimension will distort a psychophysical function. Therefore successive samples should be selected in order to spread responses evenly over the response range (Booth et al., 1983). Also, a balance of responses between excess and insufficiency of an attribute is necessary in order to avoid the misestimation of the most preferred level which commonly occurs in sensory optimisation (Conner et al., 1987). Perhaps most important of all, responses near and therefore potentially beyond the end(s) of the response dimension must be avoided. That is, in this context, samples should not be presented that would be rated totally unacceptable by an individual consumer (i.e. risk end-effects).

Such a sample-selection procedure would be expected to minimise

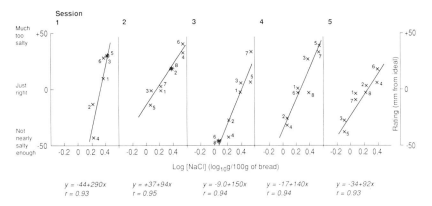

Figure 7.2 Typical unfolded acceptance triangle data from one assessor's ratings of the acceptability of bread varying in added salt level in several sessions of testing. The data show the linear relationship between acceptability ratings and salt levels scaled in equally discriminable units – log salt concentrations (redrawn from Conner *et al.*, 1988a). The numbers beside each point indicate the presentation order which was used to minimise bias within each session.

range and frequency biases and end-effects. Figure 7.2 illustrates how the bias-minimisation algorithm operates within an individual assessor (Conner *et al.*, 1988a). The key to the success of the procedure is not the layout of the response format but the selection of successive samples to suit the individual assessor's preference.

Either characterised or uncharacterised acceptance ratings from untrained individual consumers have been found to produce highly reliable linear plots against log concentrations of salt in bread (Conner *et al.*, 1988a, 1988b; Figure 7.2), sugar in a lime drinks (Conner *et al.*, 1986, 1988c; Conner, 1988), sugar in chocolate and in tomato soup (Conner *et al.*, 1988c), sugar, coffee and whitener in coffee drinks (Marie, 1987), caffeine in coffee drinks (Booth *et al.*, 1989; Booth and Conner, in press; Conner and Booth, submitted), protein in yellow fat spreads (Conner *et al.*, 1994), salt, creamer, starch and flavouring in chicken soup (Booth *et al.*, submitted a), and coffee, whitener and sweetener levels and the non-sensory factors of labelled sweetener calories in coffee drinks (Booth and Blair, 1988). Tables 7.1 and 7.2 summarise the linearity and reliability data from two major studies.

7.3   Psychophysical acceptance parameters

The straight-line psychophysical functions of acceptability responses against determining factor levels derived from the unfolded acceptance triangle can be characterised in several ways to predict choice behaviour.

Table 7.1 Linearity of blind, bias-minimised, ideal–relative saltiness ratings of samples of salt in bread (Conner et al., 1988a) and of sweetness ratings of samples of sugar in chocolate, lime drink and tomato soup (Conner et al., 1988c) by subjects within sessions

Food	Session[a]	n	Linear regression variance	
			Mean	SD
Salt in bread	1	39	0.76	0.22
	2	39	0.79	0.19
	3	39	0.80	0.13
	4	17	0.87	0.06
	5	17	0.75	0.21
Sugar in chocolate	1	18	0.58	0.28
	2	18	0.77	0.21
Sugar in lime drink	1	18	0.83	0.11
	2	18	0.88	0.08
Sugar in tomato soup	1	18	0.72	0.18
	2	18	0.71	0.19

[a] All sessions had bias-minimised presentation except session 3 of the salt study which had random-order presentation.

Table 7.2 Reliability and precision of subjects' ideal–relative saltiness ratings of salt in bread (Conner et al., 1988a) and of sweetness ratings of samples of sugar in chocolate, lime drink and tomato soup (Conner et al., 1988c). Bivariate correlation indicates the correlation between replicate samples presented in differing sessions. Variance of regression indicates proportion of variability in the ratings which can be explained by physical differences in the food samples (using data from both sessions)

Food	Sessions	Delay	n	Bivariate correlation[a]		Variance of regression[b]	
				Mean	SD	Mean	SD
Salt in bread	1–2	11 wk	39	–	–	0.72	0.17
	2–3	2 wk	39	0.91	0.15	0.76	0.11
	3–4	2 wk	17	0.91	0.17	0.77	0.09
	4–5	1 day	17	0.92	0.11	0.75	0.13
Sugar in chocolate	1–2	1 day	18	0.66	0.34	0.61	0.21
Sugar in lime drink	1–2	2 day	18	0.89	0.13	0.82	0.10
Sugar in tomato soup	1–2	1 day	18	0.83	0.15	0.65	0.16

[a] Formulations changed between sessions 1 and 2 of salt in bread study, therefore a correlation coefficient is not available.
[b] Regressions from ratings to log concentrations, taking both sessions together.
All Ps of the median F-ratios were less than 0.0001.

The unfolded acceptance triangle has the three characteristics of any linear regression equation: slope, intercept and residual variance. These three parameters can be converted to three characteristics of the mental process mediating the influence on acceptability within the individual. These characteristics represent the theoretically and practically important aspects of the individual assessor's motivation to accept the product in the tested situation: maximum preference and tolerance of off-maximum levels.

The following discussion for simplicity refers to the concentration of a constituent in relation to determining the individuals' acceptance parameters. However, any influence on acceptability – chemical, physical or conceptual – can have a most preferred level and tolerance of deviations of off-maximum levels. The only requirement is that the x-axis has units of equal discriminability.

7.3.1   Ideal point (IP)

The maximally preferred level of the varied factor is the concentration corresponding to the ideal category (e.g. 'always choose') which can be estimated by interpolation (Figure 7.1). This is the amount of the constituent at which the regression line intercepts maximum acceptance. The most preferred level is likely to be strongly related to the usually consumed level of that factor in that product for that individual (Conner, 1991a) and so the estimate of the ideal point is rather accurate. McBride and Booth (1986) found the ideal to be at least as precise an internal standard for strength of taste as an externally presented standard, possibly because ideal is so familiar. Hence, having ideal as a response category is likely to improve the precision of ratings of taste acceptability, in addition to providing a readily understandable and potentially useful parameter of acceptance.

7.3.2   Rejections ratio (RR)

The other two parameters are partly independent measures of the assessor's personal motivational characteristic of tolerance of deviations from this ideal level, i.e. the sensitivity of the psychophysical acceptance function.

One measure of tolerance of levels off-ideal is the slope of the regression line. Because the acceptance triangle, from which the line is unfolded, is in equal discrimination units, the slope is the increase in the raw rating unit per unit log increase in stimulus amount. Differences in acceptance are based on perceptual differences and these are in constant ratio of a phenomenon such as concentration (i.e. the line is on a semi-log plot). Thus, the slope is a ratio of two levels of the influence on acceptance.

It is convenient to use the assessor's rejection points as the two levels over which to measure slope. Hence, on a log scale, the slope is the rejections ratio (RR). This is the ratio of that level of the varied factor which is at the assessor's excess rejection limit (RP_{xs}) to that level of the factor which lies at the deficit rejection limit (RP_{def}) (each limit calculated by extrapolation and measured in concentration units). The rejections ratio gives a measure of the range of the varied factor that is tolerable to the individual. This measure is solely dependent on the slope of the regression line for the unfolded acceptance triangle which in turn entirely depends on the assessor's interpretation of the exact wording of the rejection response categories. In that sense, RRs are rather subjective. They may also be rather unreliable, since extreme levels of the acceptance factor which may be unfamiliar are calculated by extrapolating beyond the data.

7.3.3 Tolerance discrimination ratio (TDR)

The slope of the intensity ratings plotted against the stimulus amounts (typically in ratio, i.e. logarithmic units) has long been used in psycho-physics as the measure of supra-threshold sensitivity. (When the ratings are also plotted logarithmically, this slope is the exponent of the power function that S.S. Stevens alleged ratio instructions to yield.) With the linear semi-log plots of the unfolded acceptability triangle, the slope is the increase in rating per log increase in stimulus amount.

However, it has long been realised (Torgerson, 1958) and recently reiterated (Weiffenbach, 1989) that a more objective measure of supra-threshold sensitivity should also use the response variance (the residual variance of the linear regression in this case). In other words, discrimi-nation performance is a function of consistency, expressed as low response variability, as well as being a function of change in rating per stimulus unit (raw slope). Such an estimate of sensitivity would not necessarily depend on the stimulus levels to which the assessor has assigned the rating anchors. It would be an estimate of the discriminative power of the rating performance, free of response bias.

When responses actually are influenced by the plotted stimulus factor, the response variances should be constant over the tested range of samples: the individual assessor's data can be tested for this condition. In that situation, the change in the acceptance factor that is just discriminable in the acceptance ratings can be estimated by a version of the classical just-noticeable-difference (JND) calculation. The JND is defined as the size of increment that must be added to a standard stimulus before a sensation is aroused which is different from the standard (Torgerson, 1958) or, in terms of observables, when the response is reliably different by some suitable criterion. That is, the JND is a measure of the sensitivity of that judgment by the subject to differences in the stimuli. The JND expressed relative to the size of the standard stimulus is known as the Weber ratio: the smaller that ratio, the finer is the discrimination, i.e. the greater the sensitivity.

This measure of the JND is best stated in terms of dispersion of the subject's distribution of errors (Torgerson, 1958). Thus discrimination performance must be a function of consistency expressed as low response variability, as well as of change in rating per stimulus unit (slope). This will give a measure of the subject's ratings at distinguishing different levels of the varied factor. The traditional JND calculation can be applied to the whole psychophysical function when response variances to stimuli are equal (Booth, 1988; Conner et al., 1988a). Since the sensitivity of an acceptance response to stimulus differences may not be at the perceptual limit, we call it a just tolerable difference (JTD) to distinguish it from a

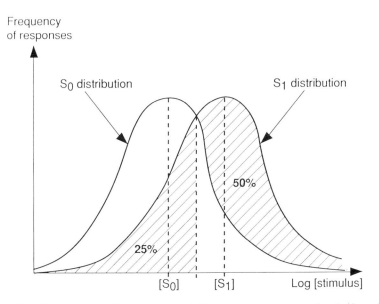

Figure 7.3 Theoretical distribution of acceptability responses to two stimuli (S_0 and S_1) varying by 1JTD in the factor determining acceptance (log [stimulus]). The responses to each stimulus are normally distributed about a particular point on the response dimension (x-axis). The stronger stimulus (S_1) produces a set of responses that overlaps with the responses to the weaker stimulus (S_0), but that on average are slightly higher. The overlap of distributions in this case is 50%. The distance between the two stimuli in response variance terms can be expressed as a z-score. For a 50% overlap of distributions as presented here, 25% of the responses to S_0 (or S_1) lie between the mean response to S_0 (or S_1) and the midpoint of the overlap. Hence the distance between the two stimuli in standard deviation units will be twice the z-score associated with that distance from the mean which would mark out 25% of the responses to that stimulus.

JND at the discriminal limit. The individual's tolerance of deviations from ideal can be calculated in determining factor concentration units using the JTD to calculate a Tolerance Discrimination Ratio (TDR), in a similar way to which the JND is used to calculate the Weber Ratio. The simplest calculation is possible when the distribution of responses to stimuli are normal, and the variances are equal and uncorrelated (Case V of Thurstone, 1927).

The derivation of the calculation is based upon the case of two stimuli, S_0 and S_1, whose average values are close enough together and their variances sufficiently large that their discriminal dispersions overlap (as in Figure 7.3). The ascending TDR (TDR_a) is taken to be the ratio of the difference between these two stimuli (S_0 and S_1) to the lower stimulus (S_0), such that, as in the traditional JND calculation, 75% of the ratings to the 'stronger' stimulus (S_1) are higher than any ratings to the 'weaker' stimulus (S_0) they could by chance be compared with. This condition occurs when the two equal distributions overlap by 50% of each distribution (Figure 7.3). This is the usually chosen condition because it is half-way between the two extremes of total indistinguishability (100% overlap) and total distinguishability (0% overlap).

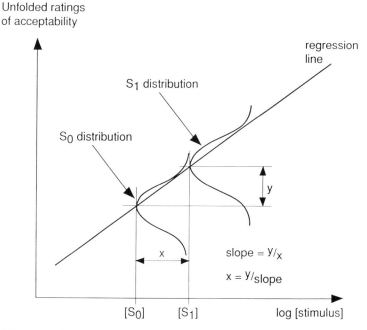

Figure 7.4 Regression between acceptability ratings and a determining factor (log[stimulus]) showing how a JTD is calculated. Assuming equal response variances, y is a simple function of the variability in responses to any stimulus level and x (JTD) is a ratio of this value to the regression slope. S_0 and S_1 are shown as separated by 1JTD.

Hence, for any two concentrations, S_0 and S_1, one JTD apart:

$$S_1 = (1 + TDR_a) \cdot S_0$$

$$TDR_a = (S_1 - S_0) / S_0 = S_1/S_0 - 1.$$

In a rating task such as used here, the ratio S_1/S_0 is obtained from the regression of log stimulus concentrations onto ratings (Figure 7.4). The 75% criterion will be met when the distributions of the responses to two stimuli overlap by 50% of either distribution. The difference between the S_1 and S_0 in standard deviation units will thus be twice the z-score associated with an area of 25% from the mean of a distribution ($z = 0.675$). In Figure 7.4, the distance y will be a function of this z-score and the mean residual response variance. The difference between the two stimuli in log concentration units (x in Figure 7.4) will then be the product of double this z-score (2×0.675) and the ratio of the response SD (square-root of the mean residual response variance about the regression line) to the slope (m) of the regression line of responses on concentrations (Conner et al., 1988a). Hence:

$$TDR_a = S_1/S_0 - 1 = \text{antilog} (\log S_1 - \log S_0) - 1$$
$$= \text{antilog} (2 \times 0.675 \times SD / m) - 1$$

There is a corresponding descending TDR (TDR_d), such that:

$$S_0 = (1 - TDR_d) \cdot S_1$$

$$TDR_d = 1 - S_0/S_1$$

and $\qquad\qquad TDR_d = 1 - 1/(1 + TDR_a)$

The conditions for applicability of this simplified calculation from all the data are equal response variances and a linear psychophysical function, both of which should be tested for before application of the calculation. According to the causal theory of the mental processes behind the acceptance triangle, these conditions are attainable by unbiased rating performance.

The TDR is likely to be higher than the Weber ratio because the assessor is unlikely to be motivated to operate at the perceptual limit. Indeed, in studies where comparisons between the TDR and Weber ratio have been made (Conner et al., 1988a) the TDR tends to be very similar to the Weber ratio or to an integer multiple of the Weber ratio. It seems that assessors use inter-category distances in their ratings which put at least two Weber ratios between end-points and small multiples of WRs when they are being less discriminating.

Note that this form of analysis uses all the data from what is sometimes called 'direct scaling' and furthermore generates from these ratings the same output as 'indirect scaling', i.e. estimates of the JNDs at all the

Table 7.3 Across panel stability of individual's acceptance parameters (IP, RR and TDR) in four different foods

Food[a]	Sessions	Delay	n	Bivariate correlation		
				IP	RR	TDR
Salt in bread	4–5	1 dy	17	0.65**	0.07	0.41*
Sugar in chocolate	1–2	1 dy	18	0.77**	0.29	0.08
Sugar in lime drink	1–2	2 dy	18	0.86**	0.73**	−0.08
Sugar in tomato soup	1–2	1 dy	18	0.67**	0.52*	−0.01

$*p < 0.05$; $**p < 0.01$.
[a] Salt in bread (Conner *et al.*, 1988a), sugar in chocolate, lime drink and tomato soup (Conner *et al.*, 1988c).

stimulus levels tested (McBride, 1983). Therefore we do not need to do laborious JND measurements, by the method of comparisons (Guilford, 1954) or whatever: we can calculate a JND from a few unbiased ratings.

In various studies (Conner *et al.*, 1988a, 1988c) we find that a consumer's TDR is sometimes much greater than the perceptual limit, as estimated by Weber's ratio of discrimination for stimulus descriptions (Schutz and Pilgrim, 1957). This indicates that some people are not pitting their ratings against the perceptual limit; that is, motivational factors are flattening the slope of the psychophysical function. Our data so far on sweetness (Conner *et al.*, 1988c) and saltiness (Conner and Booth, 1992) indicate that a majority of people are sufficiently intolerant of non-preferred magnitudes of a variety of tastants in foods to give a distribution of TDRs across assessors which is heavily skewed to a sharp low limit, somewhat above the two-alternative forced-choice Weber ratio.

These three parameters of an individual's preference (IP, RR and TDR) have been found to show a certain amount of stability across time (Table 7.3), supporting the claim that these parameters are truly individual characteristics of rating performance.

7.4 Aggregation of individuals' acceptance parameters

Once the characteristics of individual motivation (IP, RR and TDR) have been derived for the factor (or combination factor) tested, the implications for choice between real variants of a product in the situation of use assessed can be calculated for the population sampled, by aggregating characteristics across the panel by means of simple summation. The validity of the aggregate depends on the representativeness of the panel tested as it does for any consumer preference testing. However, the precision of the market response estimate is likely to be high.

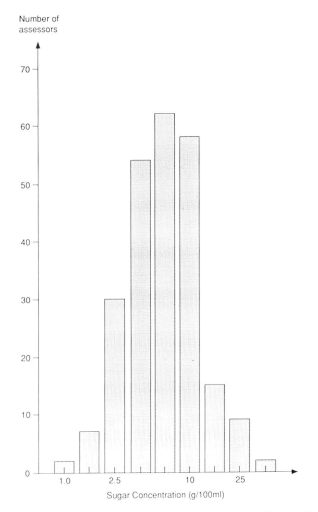

Figure 7.5 Distribution of ideal points across assessors for sugar in a lime drink (redrawn from Conner *et al.*, 1986) on a logarithmic scale of sugar concentrations.

The simplest form of summation of individuals' characteristics would be a frequency plot of ideal levels (Figure 7.5). However, such plots fail to make use of all the measures of assessors' rating performance. Two measures which do use the three characteristics of individual motivation are the ideal range and the tolerable range. The ideal range covers those levels of the factor determining acceptance that are not distinguishable from the most preferred level of that factor for that individual. This is calculated from the IP and the TDR and can be stated as:

Ideal range:

Minimum = [IP − 1JTD] to Maximum = [IP + 1JTD]

The tolerable range is those levels of the factor determining acceptance that are just tolerably different from the rejected levels for that individual. This is calculated from the rejection points (RP_{xs} and RP_{def}) and the TDR as:

Tolerable range:

Minimum = [RP_{def} + 1JTD] to Maximum = [RP_{xs} − 1JTD]

Levels of the factor within the tolerable range are just tolerably different from those levels which would be rejected. Calculations of each of these different parameters of an individual assessor's choice for sugar in a lime drink are shown in Figure 7.6.

Once the characteristics of individual motivation (IP, RR, TDR) have been derived for the combination factor or discrete factors tested, their implications for choice between real variants of a product (using either the ideal range or tolerable range) in the use situation assessed can be cumulated across the panel tested.

The IPs, RPs and TDRs have usually been found to be normally distributed across assessors against the logarithm of concentrations. This allows the cumulative percentage of assessors finding a particular factor level within their ideal or tolerable range to be summarized as straight lines on a normal probability scale. Figure 7.7 shows such a plot for a group of subjects' data on salt levels in bread.

Such plots allow us to make predictions about the likely market response to changes in the levels of the tested factor. For instance, in these salt in bread data (Figure 7.7) the ideal range lines intersect at a formulation concentration of 1.43% salt. This level was estimated to be not tolerably different from ideal (IP ± 1JTD) for 96% of these assessors and to be outside the tolerable range (RP_{def} + 1JTD to RP_{xs} − 1JTD) for virtually no one (Figure 7.7). In addition, these data indicate that a 1% salt level was in the ideal range for 53% of assessors and was below the tolerable range for only 16% of assessors. Hence a change from 1.5% to 1.0% salt would be predicted to have only a negligible effect on the market acceptability of bread. Of course such data cannot be directly converted into estimates of market responsiveness to changes in the levels of the varied factor without calibration data.

Such data are not currently available, although one study carried out did produce a recommended change in the varied factor level which was subsequently acted upon (Conner and Booth, in preparation b). Figure 7.8 shows the distribution of ideal ranges and tolerable ranges for sugar levels in this product. The standard level of sugar in the product when these tests were carried out was 8.0 g. The ideal lines on this plot intersect at 8.6 g, this level being within the ideal range for 97% of assessors. The

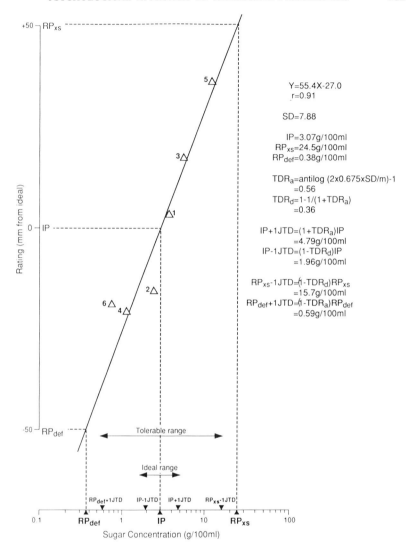

Figure 7.6 A single assessor's ratings of sweetness in a lime drink varying in added sugar level plotted on a logarithmic scale of sugar concentrations. The data illustrate the calculation of the various individual parameters of acceptance (Table 7.7). The numbers beside each point indicate the presentation order of samples indicating a bias-minimised order of presentation.

two tolerable range lines intersect at 6.5 g and this level would be outside the tolerable range for 37% of assessors. Reformulating the product from 8.0 g to 7.0 g would be predicted to have negligible effects of the numbers of assessors finding the product within their ideal range (reduced from 96% to 94%) and outside their tolerated range (decrease from 50% to

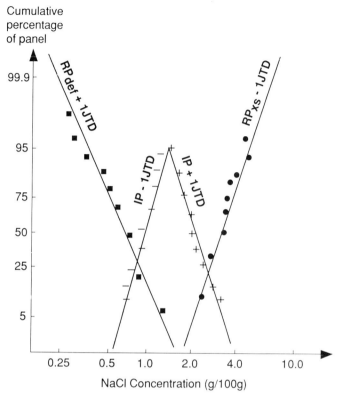

Figure 7.7 Panel acceptance and rejection lines for salt levels in bread plotted on a logarithmic scale of salt concentrations. Unfolded acceptance triangles were first fitted to each individual's data and acceptance parameters calculated. The cumulative proportion of assessors finding particular levels of salt greater than their rejection range maximum (RPxs − 1JTD: ●) or above their ideal range minimum (IP − 1JTD: −), or assessors finding particular levels of salt less than their rejection range minimum (RPdef − 1JTD: ■) or less than their ideal range maximum (IP + 1JTD: +) is plotted on the figure. Regression lines were then fitted to each cumulative plot. These data form straight lines on a normal probability scale because individuals are normally distributed on each measure against the logarithm of concentrations and so cumulative percentages plot as a straight line (redrawn from Conner *et al.*, 1988a).

41%). Implementation of this change did not result in any increases in complaints about the product. In the near future it is hoped to examine the changes in sales data for this product following this change. Changes in sales might be expected to be minor because the overall change in market acceptability was predicted to be minor. Furthermore, there is evidence to suggest that exposure to the new level of sugar would gradually become adapted to (cf. Schutz, 1991).

The exact relationship between these individual acceptance parameters and actual choice behaviour remains to be elucidated. It may be that

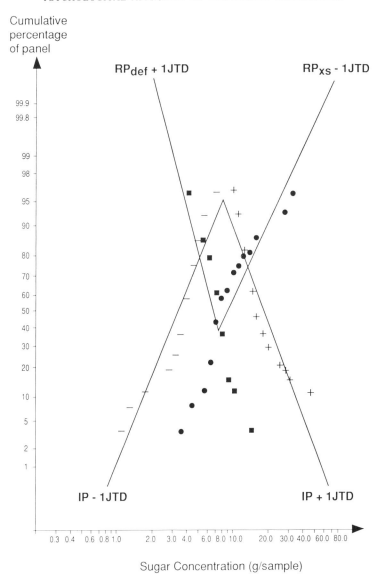

Sugar Concentration (g/sample)

Figure 7.8 Panel acceptance and rejection lines for sugar levels in a food plotted on a logarithmic scale of sugar concentrations: calculations and symbols as for Figure 7.7 (Conner and Booth, in preparation b).

different individuals use differing strategies, some rejecting products which are noticeably different from their most preferred level (i.e. outside their ideal range) and some only rejecting products outside their tolerable range. Alternatively various sorts of graded rejection are possible with greater rejection rates expected as factor levels deviate further from the most preferred level. Without the data we can only guess at the exact relationship although the ideal and tolerable ranges do appear to be plausible representations of the choices consumers make.

If acceptability is based upon familiarity as suggested above (section 7.2.1) then any formulation changes based upon these ideal and tolerated ranges (Figures 7.7 and 7.8) are likely to be adapted to in time allowing subsequent changes to be made with less loss of acceptability. A series of modest reductions in a factor in a food with time between each reduction to allow palatability to shift might be more practicable than a single large reduction because each smaller step risks losing less palatability (Booth *et al.*, 1991; Conner, 1991b; Schutz, 1991). However, experimental validation of this proposition is required. Such research is particularly important for constituents of foods which have potentially serious health consequences or for those that are undesirable for other reasons (e.g. high cost).

This individualised approach avoids the common assumption in most sensory and consumer research that the qualitative character of what is going on is the same in every head (Booth, 1988). Predictions of the proportion of the market segment accepting and rejecting any particular formulation of the product do not depend on questionable averages or assuming unrealistic consensus configurations. The fallacy of common individual structure has long been recognised in the social sciences (Robinson, 1950) and no amount of cross-tabulation of frequency data can avoid it. The fallacy is equally common in tests using preference scores or ranks and in sensory panel averaging, and indeed all analyses of grouped data aimed at diagnosis of psychological phenomena. The individualised procedure advocated here avoids the fallacy by examining causal structures within individuals and measuring the parameters of these individual causal structures. This approach thus picks up on the advantages of both idiographic and nomothetic approaches to research design and data analysis (Jaccard and Dittus, 1990).

The cumulation of individuals' responses (Figures 7.7 and 7.8) should not be interpreted as an average or typical response, as conventional response surfaces have to be. It is the total response estimated for whatever market segment the aggregated people represent, however much or little their preference structures differ. Such summaries of the data take into account variations across the panel both in individuals' most preferred levels (IPs) and their tolerances of deviations from their ideal level of the varied attribute (TDRs). Personal acceptance parameters (IP, RR and TDR) also allow predictions of the acceptability of variants not tested by the assessor, which the results of preference tests cannot do.

These calculations for the single-factor case have been programmed for PCs (Conner, 1989). This disc considerably speeds the individualised regression analyses necessary for data to be grouped into market response distributions. It can also guide individualised sample selection. As with all consumer data, such predicted market responses need calibrating against real market data. Our experience so far, however, is that action can be taken on a one-to-one interpretation of predicted against actual market behaviour at least as safely as from other types of sensory or consumer data.

7.5 Measuring determinants of acceptance in the 'real world'

All the discussion so far has been restricted to the univariate case, the effect of a single sensory factor under experimental control upon acceptance. In the real world, several factors may vary at once and interact with one another to influence acceptance, or we may not have physical measures on all the varying factors, or we may not be able to independently control the levels of each of the factors influencing acceptance, or the factors influencing acceptance may be conceptual rather than physical. The acceptance triangle theory can be applied to all these differing situations. The extra steps in the analysis in applying the acceptance triangle theory to each of these situations and initial attempts to apply the theory are described in the following sections.

7.5.1 Multiple determinants

The acceptance triangle represents a single discrete cause–effect relationship between a unitary determinant or set of determinants and the acceptance ratings. That is, the effects of several factors on acceptability can be modelled either as a single acceptance triangle with the x-axis representing some combination variable of the influencing factors or as several orthogonal (statistically independent) acceptance triangles, one representing the discrete effect of each factor.

Analysing the effects of several factors on acceptability involves two steps. The first step is the identification of a linear relationship between each orthogonal determinant and acceptability judgements. On the basis of these linear relationships, individual parameters of acceptance (IP, RR and TDR) are calculated for each identified determinant of acceptance. The second step is to determine how the various discrete determinants combine to give overall acceptability. This utilises the individual parameters of acceptability calculated for each discrete determinant. Any interactions between determinants must be accounted for at the identification stage as the second stage simply models how the discrete determinants combine to predict acceptability.

The process of identification is based upon analysing representative individual's data for linear relationships between measures on the determining factors and the (unfolded) acceptability responses. In some cases (i.e. where there are interactions between determinants) these linear relationships will only be obtained by using some form of combination of measures of the determinants. For example, the sweetness of a chocolate is likely to be a function of both added sucrose level and other factors which alter the rate at which the sucrose is dispersed in the mouth. The way in which the analysed factors are integrated should be investigated by the usual scientific strategy of testing straightforward hypotheses first. The nature of the integration can be tested within individuals using hierarchical multiple regression. The integration could well be expressed in a simple and determinate formula. The multidimensional psychophysical function should behave in the same way as a discrete function, i.e. like a single factor acceptance triangle. (Indeed, the discrete function may be complex at another level of analysis.) Once identified in representative assessors these combination functions can be applied to each individual assessor's data to calculate the parameters of acceptance (IP, RR and TDR) for each discrete determinant. This is not to underestimate the difficulty of identification which may represent a major obstacle to the application of this methodology to the multi-factor situation.

Figure 7.9 illustrates this analysis for an assessor's ratings of the overall acceptability (between 'always choose' and 'never choose') of five samples of chicken cup-a-soup varying in added salt and creamer levels (Booth and Conner, 1990). The data were first unfolded by the experimenter into ratings that are different from ideal because of excessive or insufficient levels of the determining factor. One simple coarse strategy for achieving this is to assume that factor levels are close to ideal in the most preferred sample presented. The first task in the analysis of such data is to identify each substantial influence upon a person's acceptance as a discrete factor (Figure 7.9a,b). The independent source of influence is identified by the fact that its effects upon acceptance can be observed as linear performance by the assessor. That is, a straight-line relationship can be found after unfolding of the data. Figure panels 7.9a and 7.9b identify the influences on overall acceptability of salt level and creamer level respectively.

The way in which the identified discrete influences upon acceptance combine within an individual assessor to produce an overall acceptance response is also an empirical matter to be tested on the individual's data. Booth (1987a, 1987b, 1988) reviews various plausible decision models that have been used on group data in consumer behaviour and in cognitive psychology (Anderson, 1981) and recently discussed by Mullen and Ennis (1987) in terms of multidimensional difference spaces (city block or Euclidean). As with any theoretical explanation of how a system works, more than one integration formula may fit a given set of data very well; in such a case we have to collect more data from that individual under those

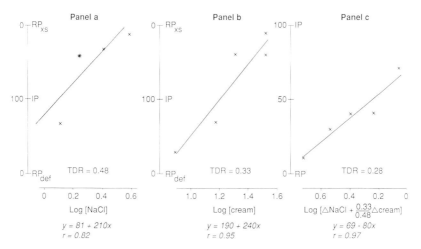

Figure 7.9 One assessor's ratings of overall acceptability for five samples of chicken cup-a-soup varying in added salt and creamer. Panel (a) shows the identification of the effect of salt level and panel (b) the identification of the effect of creamer level on overall acceptability. Panel (c) shows the fit of the overall acceptability ratings to the modelled equation of the combined effects of the salt and creamer levels (redrawn from Booth and Conner, 1990). Note that the x-axis in Panel (c) is reversed and scaled in terms of distances from the individual's ideal salt and creamer formulation of soup.

conditions in order to determine which hypothesised interaction pattern was nearest the truth. The way in which the analysed factors are integrated should be investigated by the usual strategy of testing straightforward hypotheses first.

The acceptance triangle theory specifies the fact that drops in acceptability from 'always choose' are directly related to distances from most preferred levels (IP) for each discrete (orthogonal) determinant of acceptance. Hence, we merely need to scale these distances from most preferred level. This can be achieved in a variety of ways.

Perhaps the simplest would be a first-past-the-post decision (Marie *et al.*, submitted). In this case, a particular sample would be assigned a probability of choice of one ($P = 1$) if the levels of each of the discrete determinants were within a particular range and a probability of choice of zero ($P = 0$) if any one level were outside a specified range. Plausible ranges to use might be the ideal range (IP $\pm$ 1JTD) or the tolerable range (RP_{def} + 1JTD to RP_{xs} − 1JTD) for each discrete factor (this is the model used in Figures 7.7 and 7.8). These probabilities of choice values for each sample presented to an assessor are then correlated with the actual overall acceptability ratings given by the assessor. Note that such a formula applies weighting factors to the physical input measures according to the relative sizes of the TDRs, thereby converting them into perceptual measures. Thus, this integration algebra is an empirical hypothesis about the mental mechanism by which the assessor decides what she or he wants.

A slightly more complex alternative is to use graded decisions (between $P = 1$ and $P = 0$) based upon either of the same two limits (i.e. the ideal range or the tolerable range) for samples within these ranges. In this case a probability of choice value is calculated for each discrete determinant in each sample. The value is $P = 1$ if the level of the varied factor is ideal and $P = 0$ if it is at or beyond either limit. When it is between these two limits its value is calculated as a ratio of its position between the maximum and minimum choice points to the size of this range (i.e. $P = 1 - (IP - Sample)/(IP - [RP_{def} + 1JTD])$ for the tolerable range calculations). Within each sample an overall probability of choice value can be calculated from the P values associated with the levels of each discrete determinant. Averaging across these P-values is one possibility. If we consider these probabilities of choice as distances in a multidimensional space then averaging becomes a Euclidean averaging metric of the summed distances from each $P = 1$ value. An alternative combination method within a multidimensional space would be an adding or city-block type metric. Further combinations are possible but have not been found to be widely applicable to this type of data (Conner, 1988). The similarity between this analysis and the fitting of response surfaces to individuals' acceptability data is worth noting.

A further alternative is to use the individual parameters of acceptance in directly calculating the distances in a multidimensional space. For each discrete determinant of acceptability in each sample presented to each individual assessor the number of JTDs from IP is calculated. These 'distances' can then be combined by either a Euclidean averaging or city-block metric. For instance, Figure 7.9c illustrates combination of the distances of salt and creamer from their ideal levels according to a city-block metric, i.e. adding salt distance to creamer distance to give the overall distance of acceptability from ideal. The distance of the level of a constituent in a soup sample from the assessor's most preferred level is measured in number of just tolerance differences (JTDs). The result is a good-fitting combination equation for the assessor's observed ratings of the overall acceptability of chicken cup-a-soup at the different levels of salt and creamer tested (Figure 7.9c). Note how the TDR of the combination variable (Figure 7.9c) is lower than the TDRs of its two components (Figure panels 7.9a, 9b): this greater sensitivity or causal strength is evidence that this joint percept of salt and creamer is closer to what really controls the choice ratings than is either of the discrete percepts of salt or creamer. This is a combination equation of the effects of salt and creamer on overall acceptance in this individual.

The effectiveness of these various combination rules can be tested across groups of assessors to determine the best prediction of individual ratings of overall acceptability. Booth *et al.* (1991) report that a Euclidean averaging metric of JTD distances has tended to produce the best fits to their data, accounting for 60% to 98% of the variability in individuals' overall acceptability responses to complex foods and meals.

Once the best fitting combination rule for a particular set of data has been determined, the rule can be applied to each member of the group of assessors to calculate the cumulative acceptability of particular formulations. If these calculations are performed across panels of assessors representing differing populations of interest then differing predictions can be made for differing segments of the market.

Where more than a single factor is varying it is often simpler to represent changes in acceptability as a table rather than as a plot (Figures 7.7 and 7.8). Table 7.4 illustrates such changes in acceptability across a panel of assessors for varying levels of salt, flavouring, creamer and starch in a chicken cup-a-soup (Conner *et al.*, in preparation). Individual psychophysical functions were first determined for each determinant in each assessor and then differing combination models were applied across assessors. In this case a Euclidean averaging model of distances from ideal within an individual's tolerable range was found to give the best predictions of acceptability (mean correlation between predicted and observed acceptability responses: $r = 0.78$, SD $= 0.18$).

7.5.2 Unidentified determinants

This description of the methodology has concentrated (for reasons of clarity) on readily described and varied chemical and physical determinants of acceptance. However, this approach is equally applicable to the common cases where little control is possible over variations in the influential factors or where they are hard to describe. It can still be used even when little is known about the physicochemical variations in the marketed brands (Booth, 1988).

In such cases, the group mean descriptive responses of a trained panel of assessors to each presented factor (panel scores) can act as values for a 'pseudo-physical' factor. Plots of individuals' responses against such a variable can be used to derive personal ideal points and measures of individual tolerance of deviation from ideal, expressed in terms of the panel scores. These data can then be used to derive operational information about the tested brands in terms of those brands and consumers' vocabulary.

An example of this approach for eight milk chocolates is provided in Table 7.5. In this study (Conner and Booth, in preparation a) assessors rated eight commonly available milk chocolates for overall acceptability and relative to their most preferred levels of sensory characteristics in vocabulary they generated themselves from the first sample tasted. The

Table 7.4 Panel acceptance levels for salt, creamer, chicken flavouring and starch levels in chicken cup-a-soup. Each figure represents the proportion of assessors (total $N = 76$) accepting particular levels of the four varied factors in chicken cup-a-soup. Calculations are based upon a summation of distances between ideal (IP: probability of choice $= 1$) and rejection ($RP_{xs} - 1$JTD or $RP_{def} + 1$JTD: probability of choice $= 0$) for each assessor summated across assessors (Conner et al., in preparation)

Starch level	Cream level	Chicken level 2.0					Chicken level 4.0					Chicken level 8.0				
		Salt level					Salt level					Salt level				
		0.25	0.5	1.0	2.0	4.0	0.25	0.5	1.0	2.0	4.0	0.25	0.5	1.0	2.0	4.0
2.5	2.0	9.5	13.3	19.1	20.0	15.8	12.5	16.3	22.4	23.8	18.9	9.2	12.8	18.3	19.3	15.4
	4.0	10.5	14.5	20.7	21.4	16.8	13.9	18.3	24.7	26.1	20.7	10.3	13.9	19.9	20.8	16.4
	8.0	11.7	16.1	23.2	23.7	18.3	15.4	20.1	27.6	28.8	22.5	24.3	15.4	22.2	22.9	17.9
	16.0	13.8	19.5	28.3	28.6	21.8	18.9	25.5	35.1	35.9	27.8	13.4	18.9	27.4	27.8	21.4
	32.0	12.5	17.2	24.1	23.9	18.9	16.2	21.3	28.6	29.1	23.0	25.4	16.7	23.4	23.3	18.6
6.25	2.0	13.4	17.9	25.3	26.2	20.9	18.6	23.7	32.2	33.7	27.0	13.2	17.4	24.5	25.8	21.1
	4.0	15.0	19.6	27.4	28.3	22.6	20.9	26.3	35.4	36.8	29.5	14.9	19.3	26.6	27.8	22.8
	8.0	16.3	21.4	30.3	30.9	24.5	22.6	28.6	38.8	40.0	31.8	16.2	21.1	29.5	30.4	24.5
	16.0	20.4	27.0	38.6	38.6	30.0	28.7	36.7	50.0	50.7	39.9	19.5	25.9	37.0	37.4	29.5
	32.0	17.1	22.6	31.3	31.2	24.9	22.9	29.3	39.2	39.9	31.7	16.4	21.7	30.0	30.1	24.5
15.63	2.0	10.0	14.1	20.4	20.9	16.4	13.6	17.8	24.3	25.3	20.3	22.9	13.6	33.2	20.3	16.2
	4.0	11.1	15.4	22.0	22.2	17.5	15.3	19.7	26.7	27.6	22.1	24.1	14.9	21.2	21.6	17.4
	8.0	12.2	16.8	24.5	24.5	19.1	16.7	21.7	29.6	30.3	23.9	11.8	16.2	23.6	23.7	18.7
	16.0	14.9	20.9	30.1	29.9	23.0	20.7	27.5	37.5	37.8	29.7	13.9	19.6	28.6	28.6	22.1
	32.0	13.4	18.6	25.9	25.3	20.0	17.8	23.0	30.8	30.8	24.7	12.9	17.6	24.7	24.2	19.3

Table 7.5 Panel acceptance levels for sweetness and milkiness levels in eight formulations of milk chocolate. Each figure represents the proportion of assessors (total $N = 33$) accepting each level of the two varied factors (Conner and Booth, in preparation a). Calculations are based upon a summation of probabilities of choice ($P = 1$ if sweetness and milkiness were both within the ideal range: IP − 1JTD to IP + 1JTD; and $P = 0$ if either sweetness or milkiness were both outside this range) across the group of assessors

Milkiness (distance from ideal rating)	Sweetness (distance from ideal rating)				
	−50	−25	0	+25	+50
−50	18.2	27.3	36.4	21.2	18.2
−25	24.2	36.4	42.4	36.4	18.2
0	21.2	36.4	54.5	39.4	18.2
+25	24.2	33.3	48.5	36.4	18.2
+50	18.2	21.2	30.3	24.2	15.2

terms 'sweet' and 'milky' were used by a majority of assessors and so it was simple to generate panel scores for these milk chocolates. Physical measurements were not available on the chocolates. So the mean panel scores were used as the pseudo-physical variable onto which individual assessor's sweetness and milkiness ratings were separately regressed to give 'pseudo-physical' functions. From these functions we calculated individual parameters of acceptance (IP, RP_{def}, RP_{xs} and TDR) for panel sweetness and milkiness. The parameters were then combined to give predictions of overall acceptability within individuals. Should physical measures of the tested brands become available at a later date it is still possible to reinterpret the sensory space in terms of physical differences between the products.

Table 7.5 uses a first-past-the-post ideal range model for milk chocolates varying in sweetness and milkiness. This accounted for a mean of 40% of the variance in overall acceptability ratings by individuals. Much better fits to the data may be possible with a graduated form of decision model.

7.5.3 Non-sensory determinants

In a similar way to that described above, attributed characteristics such as price, convenience, contextual appropriateness or healthfulness can be handled objectively, just as inherent characteristics are by this approach (Booth, 1988). Indeed, Booth *et al.* (1991) argue that using a similar analysis the acceptability of complete meals can be built up from assessments of the acceptability of their constituent parts. The complexity of the causal structure of a concept is as open to investigation as is the multireceptor basis of a complex sensory factor. The conceptual factor of

Table 7.6 Choice regression parameters for an individual drinker of sweet white coffee, with and without a low calorie label (from Booth and Blair, 1988)

Constituent	Label	Beta coefficient	IP	TDR
Coffee solids	None	0.36	1.2	1.47
	Lo-Cal	0.51	1.3	0.60
Whitener	None	0.77	2.7	0.33
	Lo-Cal	0.03	5.3	–[a]
Sweetener	None	0.56	4.1	0.47
	Lo-Cal	0.35	4.2	0.61

[a] Extremely large, i.e. representing negligible sensitivity of choice ratings to differences in whitener level.

brand image, or whatever, is identified by its linear influence on acceptance performance. The interaction of sensory and conceptual factors (see Booth and Blair, 1988) is also susceptible to determinate calculation for each individual, as illustrated for sensory interactions above: indeed the 'pseudo-physical' variable is a conceptual variable. The predictions for choice between specific market-mix propositions can then be summed across a panel representing the market segment of interest (Booth, 1988, 1990).

Booth and Blair (1988) present data on the effects of a 'low calorie' label on preferences for coffee drinks. Data from an illustrative consumer are given in Table 7.6. The assessor rated the acceptability of differing levels of coffee solids, whitener and sweetener either with or without a low calorie label. The effect of the label was not to change the ideal sweetness level but to increase the tolerance for variations in sweetness around this ideal (larger TDR). However, this effect was quite small in comparison to the effects of the low calorie label on whitener and coffee solid preferences. Introduction of the label shifted this assessor's tolerance of deviations from ideal from being intolerable of whitener variations to being intolerable of coffee solids variations.

The influences of other non-sensory determinants on acceptance can be measured within individuals in a similar way. For instance, the theory could be applied to the more accurate measurement of attitudinal influences upon food choice. There is already interest in the more individualised approaches to assessing attitudinal influences on food choice (Rutter and Bunce, 1989; Rosin *et al.*, 1992). Applications of the approaches advocated here would represent a fully individualised approach to measuring sensory, attitudinal and physiological influences on food choice. The manipulation of each independent influence and observation of effects on choice would allow more precise causal measurement of their relative influences upon individual food choice than the more usual correlation of variations between subjects in attitude measurement. For example we might investigate the influences of a 'healthy eating' attitude alongside sensory influences in a food or drink by the presentation of the relevant descriptive information (e.g. nutritionally balanced, wholesome, healthy eating, etc.) along with the sample varying in the sensory factors of interest (Conner, 1993). In this way both attitudinal and sensory factors could be manipulated and their influence upon acceptance measured. As in all the instances described the assessor is only required to rate the food or drink product for overall acceptability. As long as the samples are presented in such a way that the effects of the varied factors are uncorrelated, the influences of each factor can be measured in the overall acceptability ratings. This offers us the prospect of being able to investigate the full range of important influences upon individual food choices within a single methodological framework.

Table 7.7 Definitions of selected terms used in connection with the acceptability triangle theory

Term	Definition
Acceptance Triangle	Psychophysical theory relating acceptance responses to determinants (section 7.2.2)
Parameters of the acceptance triangle	
IP	The ideal point – the most preferred level of a factor determining acceptance (section 7.3.1)
RR	The rejections ratio – the ratio of that level of the factor determining acceptance that would be rejected as excessive to that level which would be rejected as insufficient (section 7.3.2)
JTD	Just tolerable difference (analogous to a just-noticeable-difference) – a measure of how well assessors discriminate differing levels of the factor determining acceptance in terms of the size of difference required to produce ratings which are reliably different (section 7.3.3)
TDR	Tolerance discrimination ratio (analogous to a Weber ratio) – a ratio of the JTD to the factor level at which it is measured, i.e. a value which is constant across the range of values of the factor determining acceptance (section 7.3.3)
Identification	Process of identifying the factor(s) determining acceptance responses (section 7.5.1)
Ideal range	A model of how a single factor influencing acceptance determines choice. Factor levels within the 'ideal' range are assigned a probability of choice between $P = 1$ and $P = 0$, levels outside this range $P = 0$ (section 7.4)
Tolerable range	A second model of how a single factor influencing acceptance determines choice, using a 'tolerable' range (section 7.4)
Decision models	Models for combining discrete determinants of acceptance in order to predict acceptability possibly using the above ideal or tolerable range models (section 7.5.1)

7.6   Conclusions

The acceptance triangle theory provides a widely applicable theory for investigating the full range of influences upon consumers' judgements of the acceptability of products. As such it includes a number of novel and unfamiliar features (Table 7.7). The theory guides both the collection and analysis of data and additionally generates precise and operational specifications for improved, differentiated or new products aimed at differing sectors of the market. At the same time, the theory offers the possibility of a greater fundamental understanding of the way in which consumers make decisions about choosing a product. This offers us the prospect of being able to investigate the full range of important influences upon individual food choices within a single methodological framework. Whilst there are a number of aspects of the theory requiring further empirical clarification, the major features of the theory make it worthy of more widespread application.

Acknowledgements

My thanks to David Booth and Hal MacFie for their helpful comments on earlier versions of this chapter.

References

Anderson, N.H. (1981) *Foundations of Information Integration*, New York, Academic Press.
Bertino, M., Beauchamp, G.K., and Engleman, K. (1982) Long-term reduction in dietary sodium alters the taste of salt. *American Journal of Clinical Nutrition*, **36**, 1134–44.
Bertino, M., Beauchamp, G.K., and Engleman, K. (1986) Increasing dietary salt alters salt taste preference. *Physiology and Behavior*, **38**, 203–13.
Birch, L.L. and Marlin, D.W. (1982) I don't like it; I never tried it: effects of exposure on two-year-old children's food preferences. *Appetite*, **3**, 353–60.
Booth, D.A. (1985) Food-conditioned eating preferences and aversions with interoceptive elements: learned appetites and satieties. *Annals of the New York Academy of Science*, **443**, 22–41.
Booth, D.A. (1987a) Individualised objective measurement of sensory and image factors in product acceptance. *Chemistry and Industry*, 441–6.
Booth, D.A. (1987b) Objective measurement of determinants of food acceptance: sensory physiological and psychosocial, in J. Solms, D.A. Booth, R.M. Pangborn, and O. Raunhardt (eds) *Food Acceptance and Nutrition*, Academic Press, London, pp. 1–24.
Booth, D.A. (1988) Practical measurement of the strengths of influence on what consumers do: scientific brand design. *Journal of the Market Research Society*, **30**, 127–46.
Booth, D.A. (1990) Designing products for individual customers, in R.L. McBride and H.J.H. MacFie (eds) *Psychological Basis of Sensory Evaluation*, Elsevier Applied Science, London, pp. 163–93.
Booth, D.A. and Blair, A.J. (1988) Objective factors in the appeal of a brand during use by the individual consumer, in D.M.H. Thomson (ed.) *Food Acceptability*, Elsevier Applied Science, London, pp. 329–46.

Booth, D.A. and Conner, M.T. (1990) Characterisation and measurement of influences on food acceptability by analysis of choice differences: theory and practice. *Food Quality and Preference*, **2**, 75–85.

Booth, D.A. and Conner, M.T. (in press) Subception of caffeine in coffee preference. *Applied Cognitive Psychology*.

Booth, D.A., Lee, M., and McAleavey, C. (1976) Acquired sensory control of satiation in man. *British Journal of Psychology*, **67**, 137–47.

Booth, D.A., Mather, P., and Fuller, J. (1982) Starch content of ordinary foods associatively conditions human appetite and satiation, indexed by intake and eating pleasantness of starch-paired flavours. *Appetite*, **3**, 163–84.

Booth, D.A., Thompson, A., and Shahedian, B. (1983) A robust, brief measure of an individual's most preferred level of salt in an ordinary foodstuff. *Appetite*, **4**, 301–12.

Booth, D.A., Conner, M.T., and Marie, S. (1987) Sweetness and food selection: measurement of sweeteners' effects on acceptance, in J. Dobbing (ed.) *Sweetness*, Springer-Verlag, London, pp. 143–58.

Booth, D.A., Conner, M.T., and Gibson, E.L. (1989) Measurement of food perception, food preference, and nutrient selection. *Annals of the New York Academy of Sciences*, **561**, 226–42.

Booth, D.A., Freeman, R.P.J., and Lahteenmaki, L. (1991) Likings for complex foods and meals. *Appetite*, **17**, 156.

Booth, D.A., Conner, M.T., and Birkett, R.J. (submitted a) Cognitive integration of habitual decision processes in the individual.

Booth, D.A., Marie, S., and Conner, M.T. (submitted b) Adult motivation to choose sweetness in a familiar beverage does not involve the congenital sweetness preference.

Bornstein, R.F. (1989) Exposure and affect: overview and meta-analysis of research, 1968–1987. *Psychological Bulletin*, **106**, 265–89.

Conner, M.T. (1988) *Measurement of the Sensory Determinants of Food Acceptance*. Unpublished doctoral dissertation, University of Birmingham, England.

Conner, M.T. (1989) *Acceptance Triangle Analysis Program*. School of Psychology, University of Birmingham, England.

Conner, M.T. (1991a) Sweetness and food selection, in J.R. Piggott and S. Marie (eds) *A Handbook of Sweetness*, Blackie, Glasgow, pp. 1–32.

Conner, M.T. (1991b) Tolerated sensory changes. *Appetite*, **17**, 155.

Conner, M.T. (1993) Individualised measurement of attitudes toward foods. *Appetite*, **20**, 235–8.

Conner, M.T. and Booth, D.A. (1992) Combined measurement of food taste and consumer preference in the individual: reliability, precision and stability data. *Journal of Food Quality*, **15**, 1–17.

Conner, M.T. and Booth, D.A. (submitted) Age differences in caffeine sensitivities.

Conner, M.T. and Booth, D.A. (in preparation a) Assessing the acceptability of eight milk chocolates.

Conner M.T. and Booth, D.A. (in preparation b) Individualised optimisation of sugar levels in a food product.

Conner, M.T., Haddon, A.V., and Booth, D.A. (1986) Very rapid, precise assessment of effects of constituent variation on product acceptability: consumer sweetness preferences in a lime drink. *Lebensmittel-Wissenschaft und Technologie*, **19**, 486–90.

Conner, M.T., Land, D.G., and Booth, D.A. (1987) Effects of stimulus range on judgments of sweetness intensity in a lime drink. *British Journal of Psychology*, **78**, 357–64.

Conner, M.T., Booth, D.A., Clifton, V.J., and Griffiths, R.P. (1988a) Individualised optimisation of the salt content of white bread for acceptability. *Journal of Food Science*, **53**, 549–54.

Conner, M.T., Booth, D.A., Clifton, V.J., and Griffiths, R.P. (1988b) Do comparisons of a food characteristic with ideal necessarily involve learning? *British Journal of Psychology*, **79**, 121–28.

Conner, M.T., Haddon, A.V., Pickering, E.S., and Booth, D.A. (1988c) Sweet tooth demonstrated: individual differences in preference for both sweet foods and foods highly sweetened. *Journal of Applied Psychology*, **73**, 275–80.

Conner, M.T., Pickering, E.S., Birkett, R.J., & Booth, D.A. (1994) Improved precision of consumer acceptability tests using an individualized attribute tolerance model. *Food Quality and Preference*, in press.

Conner, M.T., Land, D.G., and Booth, D.A. (submitted) Effects of an extreme first stimulus and of the range of stimuli on judgments of stimulus level.

Conner, M.T., Booth, D.A., and Birkett, R. (in preparation) Cognitive integration of factors influencing the acceptability of chicken soup.

Fechner, G. (1860) *Elements of psychophysics*, Translation by H.E. Adler, Holt, 1966, New York.

Guilford, J.P. (1954) *Psychometric methods*, McGraw-Hill, New York.

Harris, G. and Booth, D.A. (1987) Infant's preference for salt in food: its dependence upon recent dietary experience. *Journal of Reproductive and Infant Psychology*, **5**, 97–104.

Jaccard, J. and Dittus, P. (1990) Idiographic and nomothetic perspectives on research methods and data analysis, in C. Hendrick and M.S. Clark (eds) *Research Methods in Personality and Social Psychology*, Sage, London, pp. 312–51.

Marie, S. (1987) *Perception of Aroma From Food in the Mouth*. Unpublished doctoral dissertation, University of Birmingham, England.

Marie, S., Land, D., and Booth, D.A. (submitted) Individualised estimation of preferred constituent levels and purchase choice for a beverage.

McBride, R.L. (1983) A JND-scale/category scale convergence in taste. *Perception and Psychophysics*, **34**, 77–83.

McBride, R.L. (1985) Stimulus range influences intensity and hedonic ratings of flavour. *Appetite*, **6**, 125–31.

McBride, R.L. and Booth, D.A. (1986) Using classical psychophysics to determine ideal flavour intensity. *Journal of Food Technology*, **21**, 775–80.

Meiselman, H.R. (1984) Consumer studies of food habits, in J.R. Piggott (ed.) *Sensory Analysis of Foods*, Elsevier, London, pp. 243–304.

Moskowitz, H.R. (1982) Utilitarian benefits of magnitude estimation scaling for testing product acceptability, in J.T. Kuznicki, R.A. Johnson, and A.F. Rutkiewic (eds) *Selected Sensory Methods: Problems and Approaches to Measuring Hedonics*, American Society for Testing and Materials, Philadelphia, pp. 11–33.

Moskowitz, H.R. (1983) *Product Testing and Sensory Evaluation of Foods*. Food and Nutrition Press, Westport, CT.

Moskowtiz, H.R., Kluter, R.A., Westerling, J., and Jacobs, H.L. (1974) Sugar sweetness and pleasantness: evidence for different psychological laws. *Science*, **184**, 583–85.

Mullen, K. and Ennis, D.M. (1987) Mathematical formulation of multivariate euclidean models for discrimination methods. *Psychometrika*, **52**, 235–49.

Pangborn, R.M. (1981) Individuality in responses to sensory stimuli, in J. Solms and R.L. Hall (eds) *Criteria of Food Acceptance*, Zurich: Foster Verlag, Zurich, pp. 177–219.

Peryam, D.R. and Girardot, N.F. (1957) Advanced taste-test method. *Food Engineering*, **24**, 58–61.

Pliner, P. (1982) The effects of mere exposure on liking for edible substances. *Appetite*, **3**, 283–90.

Poulton, E.C. (1979) Models for biases in judging sensory magnitude. *Psychological Bulletin*, **86**, 777–803.

Poulton, E.C. (1987) Bias and range effects in sensory judgments. *Chemistry and Industry*, 18–22.

Poulton, E.C. (1989) *Bias in Quantifying Judgments*, Lawrence Erlbaum, London.

Riskey, D.R., Parducci, A., and Beauchamp, G.K. (1979) Effects of context in judgments of sweetness and pleasantness. *Perception and Psychophysics*, **26**, 171–6.

Robinson, W.S. (1950) Ecological correlations and the behaviour of individuals. *American Sociological Review*, **15**, 351–7.

Rosin, R., Tuorila, H., and Uutela, A. (1992) Garlic: a sensory pleasure or a social nuisance? *Appetite*, **19**, 133–43.

Rutter, D.R. and Bunce, D.J. (1989) The theory of reasoned action of Fishbein and Ajzen: a test of Towriss's amended procedure for measuring beliefs. *British Journal of Social Psychology*, **28**, 39–46.

Schutz, H.G. (1991) One small step at a time: healthier and just as nice. *Appetite*, **17**, 157.

Schutz, H.G. and Pilgrim, F.J. (1957) Differential sensitivity in gustation. *Journal of Experimental Psychology*, **54**, 41–8.

Sensory Evaluation Division of the Institute of Food Technologists (1981) Sensory evaluation guide for testing food and beverage products. *Food Technology*, **35**, 50–6.

Thurstone, L.L. (1927) A law of comparative judgment. *Psychological Review*, **34**, 273–86.

Torgerson, W.S. (1958) *Theory and Methods of Scaling*, Wiley, New York.

Weiffenbach, J.M. (1989) Assessment of chemosensory functioning in aging: subjective and objective procedures. *Annals of the New York Academy of Sciences*, **561**, 56–65.

8 Modelling food choice

R. SHEPHERD and P. SPARKS

8.1 Introduction

> He would definitely have to acquire better eating habits. Learning to cook
> vegetables would be a good place to start. It was more of developing the habit
> than anything else, if he could just get into the habit of it, of buying carrots
> and turnips and cabbage etcetera. (James Kelman, *A Disaffection*, 1989, pp.
> 267–8.)

Food choice is a primary concern to those involved in producing and
manufacturing foods since their major interest is in selling food products.
However, it is increasingly being recognised that there are important
nutritional questions related to food choice. Food choice determines
nutritional status and in so far as there are influences of diet on health
and disease (Committee on Medical Aspects of Food Policy, 1984; WHO,
1991), it is of vital importance to understand the processes by which
choices are made. In particular, only with an adequate understanding of
the reasons for people's choice of foods can we attempt to change choices
and hence influence dietary patterns in line with recommendations from
those involved in promoting health. These nutritional questions are
potentially more complex than, for example, those which involve choices
between brands of the same type of food (which are likely to be seen as
highly similar alternatives by the consumer) since they relate to choices
between types of foods: choices which may lead, for example, to diets
high or low in fat, or to diets high or low in salt content. With an
increasingly plentiful and varied food supply the issue of the reasons for
food choice becomes extremely important for the area of nutrition.

8.2 Factors influencing food choice

Food choice, like any complex human behaviour, will be influenced by
many interrelating factors. There have been a number of models proposed which seek to delineate the effects of likely influences (Yudkin,
1956; Pilgrim, 1957; Khan, 1981; Randall and Sanjur, 1981; Shepherd,
1985). These models are broadly similar although they differ in emphasis.
In general these models are not quantitative. They do not attempt to

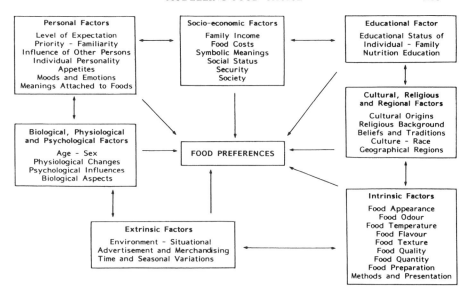

Figure 8.1 Factors influencing food preferences (from Khan, 1981; reproduced with permission CRC Press).

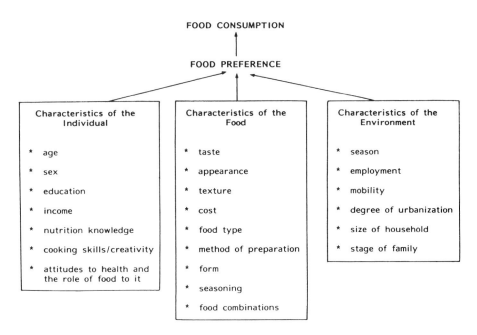

Figure 8.2 Factors influencing food preferences (from Randall and Sanjur, 1981; reproduced with permission Gordon Breach Science Publishers).

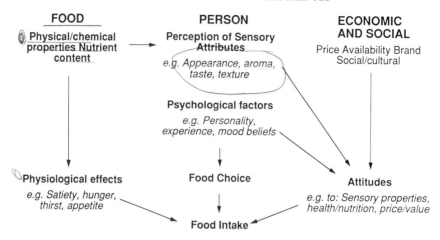

Figure 8.3 Some factors affecting food choice and intake (from Shepherd, 1985).

explain the likely mechanisms of action of the different factors, nor to quantify their relative importance or how they interact. Thus such models are really only catalogues of the likely influences. As such they can be useful in pointing to the variables to measure in studies in this area but they do not offer a framework within which to design such studies or a basis upon which to build theories of human food choice. Such models have been reviewed by Shepherd (1989).

Examples of such models are shown in Figures 8.1, 8.2 and 8.3. It can be seen that there are potentially a large number of variables. In general these variables can be divided into those related to the food, to the person making the choice and to the external environment. Authors such as Khan (1981) (Figure 8.1) and Randall and Sanjur (1981) (Figure 8.2) show these influences acting through food preferences in influencing choice. The Khan example shows all of the groups of influences as interacting in addition to affecting food preferences, thus presenting a major problem in trying to model the likely effects.

In Figure 8.3 (as in Figure 8.2) the factors are clearly categorised as related to the food, to the person making the choice and to the external economic and social environment within which the choice is made. There are also some indications of how these influences might act. Some of the chemical and physical properties of the food will be perceived by the person in terms of sensory attributes such as flavour, texture or appearance. Many aspects of the chemical composition of foods cannot be perceived and hence not every variation of chemical composition will influence choice via this route; on the other hand there will be some chemicals which are perceived at very low concentrations and can potentially have major impact. However, these sensory attributes do not, of themselves, determine whether a person will choose a food but rather the person's liking for an attribute in a particular food will be the determining factor. Liking for sensory attributes is generally food specific and probably these preferences are learned through experience (Cowart and Beauchamp, 1986). There are some common findings such as sweet foods being highly liked and bitter foods being unpalatable and there may be innate mechanisms for these preferences. However, there are many exceptions: for example, a degree of bitterness in coffee or chocolate is highly preferred. Within a particular culture there is a large degree of agreement on the appropriateness of particular sensory attributes for particular foods, but there are also very substantial differences between individuals in their preferences, which will in part lead to different food choices and diet. Understanding these individual differences in preferences and food choice is of major concern in determining the factors influencing patterns of food selection.

Other chemical components in the foods, such as the amount of protein or carbohydrate, will have effects upon the person. Thus consuming a high energy food will reduce hunger. The learning of the association between the sensory attributes of a food and its post-ingestional consequences appears to be a major mechanism by which preferences develop (Rogers and Blundell, 1990); this is not included in the model presented in Figure 8.2 but may be subsumed under the 'biological, physiological and psychological factors' heading in Figure 8.1. The converse association between ingestion and vomiting or sickness leads to a very strong aversion to the food, which is long lasting and resistant to change (Rozin, 1986).

Psychological differences between people, such as personality, may also influence food choice (Shepherd and Farleigh, 1986a, 1986b). Post-ingestional effects on psychological states, e.g. mood, sleepiness, may also influence future choice of foods through associative learning (Rogers *et al.*, 1994).

Many of the influences on food choice may be mediated by people's beliefs and attitudes. In addition to sensory preferences, beliefs about the nutritional quality and health effects of a food may be more important than the actual nutritional quality and health consequences in determining a person's choice. Likewise various marketing and economic variables may act through the attitudes and beliefs held by the person, as will social, cultural, religious or demographic factors.

8.3 Beliefs and attitudes

Thus the study of the beliefs and attitudes held by a person, and the relationship of these to the choices made, offers one means for trying to increase the knowledge of the roles played by a number of different types of factors in food choice. The assumption here is that attitudes are causally related to behaviour (McGuire, 1985). This is certainly the conceptualisation of attitude both in common use and in the area of psychology where the scientific study of attitudes originates. However, the empirical evidence for this relationship has not always been apparent.

In the nutritional literature, for example, a number of attempts have been made to relate attitudes to consumption of foods, with many studies adopting a framework known as 'knowledge–attitudes–practice' (behaviour). Here there is an explicit assumption that knowledge affects attitudes which in turn influence behaviour and therefore changes in behaviour can be brought about by increasing knowledge. What is the evidence for such a link? A meta-analysis of studies attempting to relate knowledge to attitudes and dietary intake was carried out by Axelson *et al.* (1985). Such an analysis takes the results of a number of studies and integrates them in an effort to estimate what the correlation between the variables would have been if one large study had been conducted including all of the subjects in the individual studies. Axelson *et al.* (1985) found evidence for statistically significant but small associations between these variables, the correlation between attitudes and behaviour being 0.18, and that between knowledge and behaviour even lower at only 0.10. Axelson *et al.* (1985) did not report an analysis of the relationship between knowledge and attitudes.

If we look at those studies investigating the intake of specific foods or nutrients, rather than a behaviour defined in terms of a general score for 'nutritional practices' based upon a set of behaviours (e.g. eating three meals a day), the picture is even less clear (Shepherd, 1987). In those cases where the behaviour of interest is a score for general nutritional practices there tend to be significant relationships between attitudes and behaviour (Jalso *et al.*, 1965; Schwarz, 1975; Carruth *et al.*, 1977; Foley *et al.*, 1983; Douglas and Douglas, 1984; see Eagly and Chaiken, 1993, for a discussion of criteria based upon several behaviours in the assessment of attitude–behaviour relationships). When the consumption of particular foods or nutrient intakes is the variable of interest, there tend to be only a small number of statistically significant relationships (Grotkowski and Sims, 1978; Eppright *et al.*, 1970) or indeed none at all (Perron and Endres, 1985; Werblow *et al.*, 1978).

Thus a superficial survey of this area might lead one to conclude that attitudes are not related to behaviour to an important degree, particularly when the behaviour of interest is the intake of a specific nutrient or food. The same type of finding in the social psychology field up to the 1960s led to a crisis in attitude research (Wicker, 1969), which resulted in the generation of a number of structured attitude models during the 1970s (e.g. Fishbein and Ajzen, 1975).

One example of such a structured attitude model is 'the theory of reasoned action' proposed by Fishbein and Ajzen (1975; Ajzen and Fishbein, 1980), which has been widely applied in the area of social psychology and still influences a significant amount of research work in the attitudes literature (see Tesser and Shaffer, 1990, for a view of attitude research for the 3 years up to 1990). The theory offers an essentially 'rational' account of attitude formation where the potential advantages and disadvantages of behavioural outcomes are carefully weighted by the subjective probabilities of the occurrence of those outcomes. While such an 'analytic' model of attitudes may be discordant with those orientations which emphasize 'non-rational' elements in choice (for different perspectives on this theme, see e.g. McAlister *et al.*, 1979; Zajonc, 1980; Sparks and Shepherd, 1992), it does propose the possibility of systematically identifying the influential beliefs and values in choice behaviour.

The theory of reasoned action seeks to explain rational behaviour which is volitional (i.e. behaviour which is under the control of the individual). It will thus not apply to behaviours which are not under the individual's control. With volitional behaviours it is argued that intention to perform a behaviour is the best single predictor of behaviour. Intention, in turn, is predicted by two components: the person's own attitude (e.g. whether the person sees the behaviour as good, beneficial, pleasant, etc.) and perceived social pressure to behave in this way (termed the subjective norm). These relationships are shown schematically in Figure 8.4.

$$\text{Behaviour} = \text{Behavioural intention}$$

$$\text{Behavioural intention} = w_1 \times \text{Attitude} + w_2 \times \text{Subjective norm}$$

The relative weightings (w_1 and w_2) of the attitude and subjective norm are derived from a multiple regression of these variables against behavioural intention.

The attitude is predicted by the sum of products of a set of beliefs (b_i) about outcomes of the behaviour and the evaluation (e_i) of these outcomes as good or bad:

$$\text{Attitude} = \Sigma\ b_i \times e_i$$

The subjective norm is predicted by the sum of products of normative beliefs (NB_j), which are perceived pressure from specific people or groups (e.g. doctors, family) and the person's motivation to comply (Mc_j) with the wishes of these people or groups.

$$\text{Subjective norm} = \Sigma\ NBj \times Mc_j$$

In order to develop a standard questionnaire to give to a sample of people, a set of modal beliefs can be generated from short interviews with

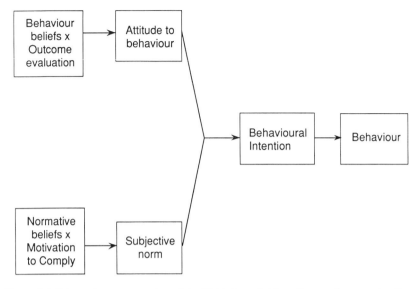

Figure 8.4 Schematic representation of the Fishbein and Ajzen theory of reasoned action.

a small number of representative individuals. Typically, the most commonly elicited beliefs are included in the questionnaire (see Ajzen and Fishbein, 1980).

One point made by the authors of this model is that influences other than beliefs, attitudes, social pressure and intention should act through these variables. Thus demographic variables, such as age or social class, should influence behaviour only through these model variables and not act as independent influences on behaviour.

As well as having been widely applied to many issues in social psychology (Ajzen and Fishbein, 1980), this model has also more recently been applied to a range of food choice issues (Axelson *et al.*, 1983; Shepherd and Stockley, 1985, 1987; Shepherd and Farleigh, 1986a, 1986b; Tuorila, 1987; Shepherd, 1988, 1989). Strong relationships between food choice and attitudes have been shown using this methodology (Shepherd, 1989). It has shown clear relationships between the components of the model with good prediction of food choice.

In these studies, the consumer's own attitude has been found to be far more important than perceived social pressure (i.e. w_1 greater than w_2 in the multiple regression analyses). The importance of different types of beliefs has been assessed by examining those belief-evaluations which differentiate those people intending to perform the behaviour and those intending not to perform the behaviour, or alternatively as the belief-evaluations best predicting attitudes. In general, beliefs concerning the taste and flavour of foods have been the most important in determining the selection of foods. One exception to this is the finding in a study of low-fat milk consumption, where belief-evaluations related to the nutritional benefits of the milks were the most important influence on consumption, although sensory and suitability aspects were also important (Shepherd, 1988).

Using this model in the area of general consumer choice, not specifically related to foods, Sheppard *et al.* (1988) carried out a meta-analysis of 87 studies (involving in total around 12 000 subjects) which met certain criteria for inclusion. They found an estimated correlation of 0.53 between intention and behaviour and a multiple correlation of 0.66 between attitude plus subjective norm against intention (Sheppard *et al.*, 1988). Thus this model has validity in the study both of general consumer choice and specifically of food choice.

As an example of the application of this model to food choice, the results of a study of fat intake will now be presented, followed by two studies extending the basic model to include other variables of interest.

8.4 Application of the theory of reasoned action to food choice

8.4.1 Fat intake: the roles of attitudes and nutritional knowledge

The study of fat intake followed a number of previous studies applying this type of model to try to understand the reasons for fat intake (Shepherd and Stockley, 1985, 1987). The basic theory of reasoned action was used to develop a questionnaire and this was completed together with a validated nutritional knowledge questionnaire (Towler and Shepherd, 1990). The study involved 538 subjects.

The nutritional knowledge questionnaire had been developed in previous trials and had been validated with groups expected to vary in nutritional knowledge (Towler and Shepherd, 1990). It contained sections on the nutrient densities of protein, carbohydrate, fat and fibre in foods, along with a set of multiple choice questions. For each of the nutrients, subjects were given a list of 20 foods and asked to indicate the 10 highest in the nutrient on an equal weight basis. They received a score of +1 for each correct answer and −1 for each incorrect answer. The multiple choice questions were more general and some included more than one correct answer.

The attitude questionnaire investigated the consumption of four types of foods (meat, meat products, dairy products and fried foods), which together contribute about 70% of the total intake of dietary fat (Shepherd and Stockley, 1985). The form of the questionnaire followed the theory of reasoned action closely but the subjective norm and normative beliefs were not included since previous studies had indicated that these were not important components in relation to the consumption of these foods (Shepherd and Stockley, 1985, 1987).

Behaviour was not measured directly but was included as self-reported frequency of consuming the food types. Intention was assessed with one question for each food type on the person's intention to eat the food during the next week. There were three attitude items for each food type, worded, for example, 'My eating meat is good', with 7-point response scales ranging from 'strongly agree' to 'strongly disagree'. The other attitude questions related to judgements that eating the food was pleasant and beneficial.

Belief items were generated in short structured interviews with 34 subjects, who did not take part in the main study. There were six beliefs for each food type. Three belief items were common to all food types: that the food was seen as healthy, high in fat and tasted good. The other three items for each food type depended upon which beliefs were found to be salient in the pre-interviews. There was a corresponding outcome evaluation item for each belief: for example 'Food which is high in fat is desirable', with response options from 'strongly agree' to 'strongly disagree', again on 7-point category scales.

The correlations between the components of the model are shown in Table 8.1. These were found to be reasonably high and were equivalent to previous findings using this model (Shepherd, 1989).

There were significant negative correlations between both the total nutritional knowledge score and the knowledge score for fat content of foods with Σ belief $\times$ evaluation, attitude, intention and behaviour scores; these were statistically significant mainly for the meat and meat products sections of the attitudes questionnaire. The correlations were negative, with those subjects with higher knowledge having, for example, a more negative attitude to the consumption of these foods. However, although a number of the correlations were statistically different from zero, they were generally small (less than $r = 0.22$), which is equivalent to those found in previous studies (Axelson et al., 1985).

Table 8.1 Correlation coefficients, for four food types contributing to fat intake, between components of the Fishbein and Ajzen theory of reasoned action. All significant at $p < 0.001$

Food types	df	$\Sigma b \times e$ versus attitude	Attitude versus intention	Intention versus behaviour
Meat	502	0.70	0.64	0.73
Meat products	501	0.63	0.59	0.78
Dairy products	515	0.58	0.40	0.57
Fried foods	509	0.59	0.61	0.77

Table 8.2 Results of regressions of attitude on to individual belief-evaluation items. The standardised (beta) coefficients and t values are shown

Belief	Regression coefficient (β)	t
Meat		(df = 497)
Healthy	0.25	7.7***
Fat	0.12	4.1***
Taste	0.53	16.9***
Expense	0.04	1.2
Protein	0.12	4.1***
Vitamins	0.07	2.2*
Meat products		(df = 498)
Healthy	0.27	7.8***
Fat	0.12	3.7***
Taste	0.51	15.6***
Expense	−0.01	−0.5
Convenience	−0.01	−0.3
Additives	0.16	4.8***
Dairy products		(df = 512)
Healthy	0.42	12.2***
Fat	0.18	5.2***
Taste	0.31	9.2***
Protein	0.16	4.6***
Vitamins	0.09	2.8**
Calories	−0.03	−0.9
Fried foods		(df = 507)
Healthy	0.16	4.6***
Fat	0.13	3.7***
Taste	0.51	14.0***
Convenience	0.01	0.3
Greasy	0.15	4.2***
Smell	−0.03	−0.9

$^*p < 0.05$; $^{**}p < 0.01$; $^{***}p < 0.001$.

The relative importance of the different belief-evaluation combinations in determining attitudes was investigated using regressions of the individual belief-evaluation scores against the attitude scores; the results of these regressions are shown in Table 8.2. In most cases, the belief-evaluation score relating to taste had the highest beta coefficient from the regression with attitude, with foods perceived as healthy second highest. For dairy products the belief-evaluation score for healthiness was higher than the one for taste. The belief-evaluation score for fat content had a lower regression coefficient than the ones for taste or health for all of the food types. Some of the other belief-evaluation scores included were also significant in the regression equations: in particular those related to protein and vitamin content, but also those referring to additives in meat products and to the greasy taste of fried foods. Belief-evaluation scores related to expense and convenience were not significant contributors to these equations.

Thus this study shows the utility of this type of model in relation to food choice. There were clear relationships between belief-evaluations, attitude, intention and self-reported behaviour. Nutritional knowledge, however, was related to these variables only to a limited degree. Other factors, such as those represented by the belief items included in the questionnaire, are likely to play a role in determining the consumption of these foods in addition to general knowledge of nutrition. In this study both the taste and health belief items were the most important, with other items like expense and convenience less so.

8.5   Extensions of the theory of reasoned action

Although the theory of reasoned action has proved successful in many applications in the food choice area, the model can be modified and extended in a number of ways. In the attitude literature there have been a number of proposed extensions. Our description here of these will be confined to those which have been empirically assessed in the domain of food choice.

8.5.1   Perceived control

One major extension of the theory of reasoned action has been proposed by Ajzen (1988) through the inclusion of a component of perceived control such that this component complements attitude and subjective norm in the prediction of behavioural intentions. Ajzen argues for the applicability of this new 'theory of planned behaviour' to non-volitional behaviours, goals and outcomes which are understood to be not entirely under the control of the person. There has been some empirical evidence in favour of including this component (e.g. Ajzen and Madden, 1986; Beale and Manstead, 1991), although the modification has not been without criticism (Fishbein and Stasson, 1990). It should be borne in mind that a lack of effect for perceived control does not necessarily present a problem for the theory of planned behaviour since its applicability is restricted to situations where problems of control exist; in those cases where perceived control is not important (i.e. where behaviour is volitional) the theory of reasoned action will apply (Ajzen, 1988, p. 136).

Perceived control has been shown to be important where the dependent variable is clearly not a volitional behaviour. Schifter and Ajzen (1985) studied weight loss, which is an outcome rather than a behaviour, and demonstrated that perceived control added significantly to the effects of attitudes and subjective norm in predicting intention and added to intention in predicting actual weight loss. Ajzen and Madden (1986) found perceived control was important in predicting an outcome (grades obtained by college students), but did not help to predict a behaviour (class attendance), although it increased prediction of intention for both. Ajzen and Timko (1986), however, investigated health behaviours (including some related to diet, e.g. eating fresh fruit, avoiding salt and cholesterol) and in some cases perceived control was more highly correlated with behaviour than were attitudes. Fishbein and Stasson (1990) found no relationship between perceived control and behaviour (attendance at a training session for operating a new telephone system). Thus, inclusion of a measure of perceived control may be important in predicting the attainment of goals and outcomes in the context of non-volitional behaviours: hence for certain choices of foods, perceived control may be a variable worthy of attention.

8.5.2   Habit

John Dewey once remarked that 'Man is a creature of habit, not of reason nor yet of instinct' (1922, p. 122) and William James, writing at the end of the last century, described habit as 'the enormous fly-wheel of society, its most precious conservative agent' (1983, p. 125). In the recent psychological writing on attitudes, habit has reappeared in the literature as an important factor influencing behaviour. Habit is considered likely to play an important role in predicting frequently performed behaviours (Ronis et al., 1989) and hence, like perceived control, its importance is likely to vary as a function of different types of behaviours. It may not play a major role in the initiation of a particular behaviour, where attitudes are more likely to be the important determinants, but when a behaviour is established it may be that habit is a key factor in the maintenance of a frequently repeated behaviour (Ronis et al., 1989). Ronis et al. suggest that habit may serve as a barrier to change but that attitudes will be more important in attempts to change behaviour. However, the apparent simplicity of this suggestion belies the further complicating factor that attitudes themselves, like behaviours, feelings and thoughts, are likely to be susceptible to the automaticity of habit (cf. Carpenter, cited in James, 1983, p. 116). Since consumptions of common foods are frequently performed behaviours, habit may be particularly important in the context of food choice (Tuorila and Pangborn, 1988).

The concept of behavioural habit is multi-faceted. It may be conceived of as the frequency of past behaviour or alternatively as behaviour that is in some sense automatic or out of the awareness of the subject (Ronis et al., 1989). Triandis (1977, 1980) developed a model of behaviour different from the theory of reasoned action, but having some features in common. In that model, in particular, habit, in addition to intention, is used in predicting behaviour. Although Triandis (1980) argues that the key feature of habit is that it is not consciously controlled, the measures he suggests relate to frequency of past behaviour rather than incorporating any notion of controllability. A number of authors have tested the inclusion in the theory of reasoned action of a variety of habit items, including measures relating to performing actions 'out of habit' (e.g. Tuorila and Pangborn, 1988), 'out of force of habit' (e.g. Wittenbraker et al., 1983) or 'without awareness' (e.g. Mittal, 1988).

In the food context, habit has previously been shown to be significantly related to the consumption of sweet, salty and fatty foods (Tuorila and Pangborn, 1988) and to coffee consumption, when habit was measured using 'out of habit' type questions. With habit defined in terms of frequency of behaviour, it did not increase prediction of visits to fast-food restaurants over that of intention alone (Brinberg and Durand, 1983), but it was found to predict meat intake and (in one analysis) salt intake over and above the effect of intention (Feldman and Mayhew, 1984). There is also evidence for the role of habit for behaviours such as seat belt use. Mittal (1988), using multiple regression analyses, found habit to add an independent effect over and above intention in predicting behaviour.

8.5.3   Self-identity

One recent suggested modification to the theory of reasoned action is that a person's self-identity may influence their behaviour independently of their attitudes towards the behaviour. Biddle et al. (1987) and Charng et al. (1988) found measures of self-identity to relate to intention independently of attitudes. Evidence for the effect of self-identity on behaviour independent of behavioural intentions is provided by Granberg and Holmberg (1990). Consideration of these three studies is included in a recent discussion of the role of self-identity within the structure of the 'theory of reasoned action' by Eagly and Chaiken (1993).

Biddle et al. (1987), in a study of college retention decisions, suggest that 'self-referent identity labelling' has an effect on behaviour that is independent of individual preferences. They found that students' ratings of what sort of person they thought they were contributed to intentions independently of the contribution of preferences. Charng et al. (1988) suggest that repeated behaviours influence people's self-concepts which then become important to those people and that carrying out those behaviours then 'conveys meaning over and above the positive or negative attitudes we may hold toward performing the behaviour itself' (p. 304). They propose that role identity as a blood donor may influence intentions to donate blood independently of attitudes towards blood donation. They demonstrate that regressions of intentions to donate blood on attitudes and role identity reveal an independent effect for each of the two predictor variables.

These findings are surprising in the context of research on the theory of reasoned action. Although a person's self-identity is conceptually distinct from that person's evaluative attitudes, and there is likely to be a bidirectional causal link between self-concept and attitudinal evaluations, it seems unlikely that there will be a *causal link* from the self-concept to behavioural intentions that is independent of the effect of those attitudinal evaluations. Thus if a person identifies themselves as a blood donor, as in the study of Charng *et al.* (1988), this would be expected to be reflected in favourable attitudes towards blood donation, so long as the behaviour were voluntary and not coerced or otherwise associated with control problems.

There are, however, some difficulties in interpreting the results of the studies by Charng *et al.* (1988) and Biddle *et al.* (1987). The attitude measures used in both studies were not the standard measures suggested by Ajzen and Fishbein (1980) and this might account for the observed independent effects of self-identity on intention. Any truly convincing demonstration of the effect of self-identity requires that the variables have been measured in the most effective way.

8.5.4 Consumption of chips

As a preliminary examination of two of the possible extensions of the basic model (perceived control and habit), a study was carried out on the consumption of chips.

Two hundred and eighty-eight subjects completed the questionnaire. This contained questions relating to the full theory of reasoned action, including the subjective norm and normative belief components. Two questions measured whether the behaviour was considered to be habitual. The first question was 'On average, how often do you eat chips out of force of habit?'. Subjects responded using a 7-point frequency scale from 'never' to 'once a day'. The second habit question had the wording 'I eat chips out of habit', with responses from 'strongly agree' to 'strongly disagree'. There was a correlation between the two habit items of 0.53 (df = 286, $p < 0.001$). Perceived control was measured using one question: 'If you were to try, what is the likelihood that you will reduce the total intake of chips in your diet?', with a response scale from 'not at all' to 'very much so'.

The correlations between the components of the theory of reasoned action are shown in Figure 8.5, along with the correlations between perceived control and habit and the measures of intention and behaviour. The standard theory of reasoned action showed reasonably good correla-

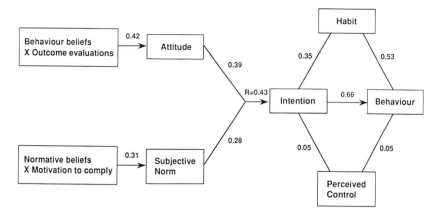

Figure 8.5 Correlations between the components of the extended theory of reasoned action for chip consumption.

tions between the components, with a multiple correlation of $R = 0.43$ for attitude and subjective norm against intention. Habit was also quite highly correlated with both reported consumption and with intention. Perceived control, on the other hand, did not correlate with either of these.

Multiple regressions were used to determine whether the extra components of habit and perceived control significantly improved the prediction of intention over and above the contribution of the standard Fishbein and Ajzen parameters of attitude and subjective norm (Table 8.3). When habit was added into this regression it showed a significant increase in the multiple correlation, but adding in perceived control failed to increase it further. Also shown in Table 8.3 are the results of a similar multiple regression predicting self-reported behaviour. Again habit showed a significant increase in the correlation but adding in perceived control had no further effect. There were no significant effects for perceived control regardless of the order in which the variables were entered into the regressions.

There was good evidence in this study for the role of habit significantly adding to the prediction of intention over and above the effect of attitude and subjective norm, and to the prediction of behaviour from intention alone. This has important consequences for the application of this model. Performing a behaviour automatically may also imply some degree of lack of personal control over the behaviour; without conscious decision making, the person may feel less in control of the behaviour. However, in the present study, perceived control did not relate to intention or to behaviour.

One major problem with this measure was the lack of variance in the

Table 8.3 Multiple correlations from study of chip consumption. Intention is predicted by the standard Fishbein and Ajzen variables of attitude and subjective norm, with habit and perceived control then added in. Behaviour is predicted by intention, with habit and perceived control then added in

Variables entered	R
Intention	
Attitude and subjective norm	0.43***
plus habit	0.51***
plus perceived control	0.52
Behaviour	
Intention	0.66***
plus habit	0.73***
plus perceived control	0.73

*** $p < 0.001$ for increase in multiple correlation.

perceived control measure in this study. Almost two-thirds of all subjects responded in categories 6 and 7, indicating that they felt they had very good control over changing their consumption of chips. Since the behaviour appears to be perceived as entirely volitional, inclusion of perceived control will not increase the prediction of intention and behaviour (Ajzen, 1988).

The conceptualisation of habit requires much more attention in future research since it is a multi-faceted notion. As a consequence of this, it becomes very difficult to construct unambiguous empirical measures (Eagly and Chaiken, 1993). The measure used in the study reported here – a self-report of performing the behaviour 'out of habit' – leaves open to subjects the interpretation of the meaning of 'habit'. However, this type of rating usually aims to incorporate the idea that the behaviour is partly automatic (and possibly out of conscious control) rather than simply being a frequently performed behaviour. However, Bargh (1989) has argued convincingly that the 'automatic–controlled dichotomy' is in itself spurious. The relative importance of, and inter-relationships between, repeated behaviour, automatic behaviour and uncontrollable behaviour needs to be addressed in future research into the concept of habit.

8.5.5 Biscuits and bread

In a further study of the role of perceived control in the prediction of food consumption, 173 subjects answered questions about their beliefs, evaluations and attitudes concerning the consumption of wholemeal bread and of sweet biscuits (Sparks and Shepherd, 1992). Subjects were also asked about their perception of control concerning the consumption of the two food types through the questions 'For me to eat the amount of X that I would prefer to eat is . . .' (with the response scale labelled 'extremely difficult' to 'extremely easy') and 'How much control do you feel you have over how much X you eat?' (with the response scale labelled 'complete control' to 'very little control'). Subjects were also asked about the frequency of specific problems (viz. about whether their consumption of the foods was affected by their being costly to buy, difficult to obtain, difficult to resist eating and whether their consumption was affected by preferences of other family members for different biscuits/bread). In the case of wholemeal bread, the addition of a measure of perceived control did not improve the prediction of behavioural intentions over and above the contribution provided by attitudes. However, in the case of biscuit consumption, the measure of perceived control did produce such an independent effect (see Table 8.4).

This finding indicates that the theory of planned behaviour will indeed apply to consumption behaviour, at least for certain foodstuffs and/or contexts of food consumption. Two important questions then become of interest for those involved with the promotion of healthy nutrition: first, which facets of the total diet are associated by consumers with control problems which in turn affect their dietary patterns; secondly, what is the nature of those control problems. In the study of wholemeal bread consumption, there was a significant relationship between increased control difficulties and both reported problems of cost and others' differing preferences (Figure 8.6). However, in the case of biscuits, the best predictor of perceived control was subjects' reports of problems of restraining themselves from eating that food.

From this study we can conclude that the modification of the theory of

Table 8.4 Regressions of intentions on attitudes, perceived control and others' attitudes for wholemeal bread and biscuits

	Regression coefficient (β)	t
Wholemeal bread		(df = 167)
Attitude	0.51	6.93***
Control	0.02	0.31
Others' attitudes	0.06	0.93
Biscuits		(df = 164)
Attitude	0.47	6.85***
Control	−0.15	−2.23*
Others' attitudes	0.07	0.99

$*p < 0.05; ***p < 0.001.$

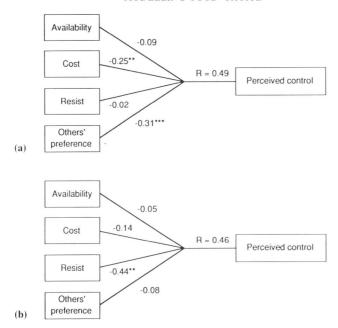

Figure 8.6 Beta regression coefficients from a multiple regression of perceived control on specific control problems for (a) wholemeal bread and (b) biscuit consumption. $**p < 0.01$, $***p < 0.001$.

reasoned action in the form of the theory of planned behaviour is likely to be an important predictive tool in certain areas of food consumption. Moreover, control difficulties are likely to be associated with a variety of types of problems which need to be addressed if dietary practices are to be successfully modified.

8.5.6   Organic vegetables

Perceived control was further investigated in a study examining the consumption of organic vegetables. The main focus of this work, however, was on the role of self-identity: whether subjects saw themselves as 'green consumers' or as concerned about 'green' issues.

261 subjects returned postal questionnaires. The questionnaire incorporated the components of the theory of reasoned action, in addition to measures of perceived control and self-identity.

Perceived control was measured with three items: (i) 'How much control do you have over whether you do or do not eat organic vegetables?', with the response scale end-points marked 'very little control' and 'complete control'; (ii) 'For me to eat organic vegetables is . . .', with response scale endpoints marked 'extremely difficult' and 'extremely easy'; (iii) 'If I wanted to, I could easily eat organic vegetables whenever I eat vegetables', with response scale endpoints marked 'extremely unlikely' and 'extremely likely'. Item scores were summed for the overall measure of 'perceived control'.

Two measures of identification with green consumerism were constructed. The first consisted of the statement 'I think of myself as a green consumer' and the second consisted of the statement 'I think of myself as someone who is very concerned with green issues' (response scales were marked 'disagree very strongly' and 'agree very strongly'). These two measures were summed to create the measure of 'green identity'.

The basic structure of the theory of reasoned action/planned behaviour was corroborated by correlational analyses: the summed products of beliefs and evaluations ($\Sigma\ b \times e$) correlated significantly with attitudes ($r = 0.45$, df $= 207$, $p < 0.001$), attitudes correlated significantly with intentions to eat organic vegetables the following week ($r = 0.37$, df $= 207$, $p < 0.001$), as did subjective norm ($r = 0.29$, df $= 207$, $p < 0.001$). 'Green' identity correlated significantly with attitudes ($r = 0.43$, df $= 207$, $p < 0.001$), with subjective norm ($r = 0.24$, df $= 207$, $p < 0.05$) but not with perceived control ($r = 0.01$, df $= 207$, ns).

To test the hypothesis that 'green identity' would *not* add to the prediction of intentions over and above the contributions provided by attitudes, subjective norm and perceived control, intentions to eat organically-produced vegetables the following week were regressed on attitudes, subjective norm, perceived control and green identity. As expected, attitudes ($\beta = 0.21$, $p < 0.01$), subjective norm ($\beta = 0.16$, $p < 0.05$) and perceived control ($\beta = 0.26$, $p < 0.001$) all revealed independent effects. However, a highly significant effect was also apparent for 'green identity' ($\beta = 0.22$, $p < 0.01$) (Figure 8.7).

When perceived control was regressed on the four control problems, there was a significant effect for the problem of lack of availability in the shops ($\beta = -0.18$, df $= 224$, $p < 0.05$), but no effects for family influence ($\beta = -0.11$, df $= 224$, ns), friends' influence ($\beta = 0.13$, df $= 224$, ns) or cost ($\beta = -0.03$, df $= 224$, ns). However, only 5% of the variance in perceived control was explained by these measures.

The findings of this study refuted the hypothesis that measures of self-identification with green consumerism would not lend independent contribution to the prediction of intentions over and above the contributions of attitudes. The expectation that self-identity would only effect its influence through evaluative attitudes was shown to be false. It does seem therefore that this study lends further support to the proposal that the theories of reasoned action and planned behaviour need to take serious account of self-identity and, with it, of behavioural outcomes that take the form of psychological effects (such as the strengthening or 'affirma-

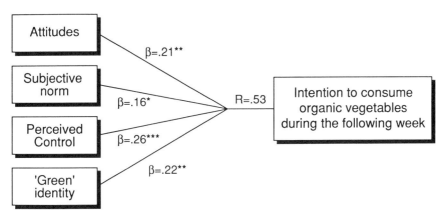

Figure 8.7 Beta regression coefficients from a multiple regression of intentions to buy organic vegetables on to attitude, subjective norm, perceived control and self-identity. $*p < 0.05$, $**p < 0.01$, $***p < 0.001$.

tion' of self-identities). Such outcomes, we have to concede, 'may not be especially salient when respondents rate behaviours on the evaluative scales used to assess attitudes toward the act' (Eagly and Chaiken, 1993).

There is now a body of empirical studies indicating the importance of self-identity in the overall structure of the theory of reasoned action. What is now required is closer conceptual attention to the issue with the aim of providing an account of the psychological processes involved whereby the influence of self-identity is realised. It is likely that the contribution of more sophisticated modelling techniques (e.g. structural equation modelling) will be of use here but it is important that this is combined with a feasible psychological account of influential processes.

8.6 Conclusions

In order to construct more general theories of how behavior is affected by attitudes toward targets and toward behaviors, social scientists must move beyond simple, volitional behavior and, to do this, must place attitudes within a theoretical structure that includes the major nonattitudinal determinants of behavior (e.g. habits, self-identity, norms). (Eagly and Chaiken, 1993.)

Food choice is influenced by a large range of potential factors. Many models put forward in this area involve merely listing the likely influences rather than offering a framework for empirical research and practical application. Although there is general agreement on the types of influences likely to be important, the integration of these factors into a coherent and quantitative model of food choice remains an area in need of much development.

The attempt to model food choice via an understanding of people's beliefs and attitudes requires a structured framework within which to measure and relate the variables of interest. One model from social psychology for achieving this is the theory of reasoned action developed by Fishbein and Ajzen. This model generally reveals good prediction of behaviour and can be used to determine the relative importance of different factors in influencing food choice through the hypothesised structured relationships between belief-evaluations and attitude and between attitude-subjective norm and intentions. This model has been successfully applied to the choice of a wide range of foods.

There are, however, a number of possible modifications and extensions to this model which need to be further developed and tested. Some of these extensions have been described in this chapter. In particular, the influential role of perceived control has been indicated. Habit, also, is a multi-faceted concept which requires more careful conceptual and empirical attention. Self-identity has also surprisingly shown effects independent of attitude and subjective norm and the role of this variable in determining behaviour requires more detailed research in the future. A consideration of moral obligation has also proved to be useful in some studies outside the food area (see Eagly and Chaiken, 1993) and may be particularly important where food choice is made on behalf of others or where sensitive political and social issues are at stake (e.g. biotechnological/genetic engineering processes in food production).

Concurrent with this interest in modifications to the basic structure put forward by Fishbein and Ajzen and a need to maintain a coherent account of the psychological processes at work in the behaviour that is under investigation, methodological problems need to be examined and potentially useful methodological procedures explored. A number of measurement issues need to be addressed, ranging from the need for improved operationalisation of theoretical constructs, through more sophisticated and sensitive methods of eliciting beliefs, such as repertory grid techniques (e.g. McEwan and Thomson, 1988), to a consideration of some of the pitfalls and problems of scale measurement and data analysis (e.g. Sparks et al., 1991).

Furthermore, the application of within-subjects' assessment of different choice alternatives using the theory of reasoned action has yet to be exploited in the food choice area, even though in other areas of research, such within-subjects designs provide impressive results (e.g. Davidson and Morrison, 1983). The present authors and their colleagues are also currently undertaking work on this issue within the domain of food choice.

The Fishbein and Ajzen model has been successfully applied to a wide range of food choice situations. It is likely that modifications to the theory will lead to improved statistical predictions of behaviour. At present it is unclear as to the extent to which the theory's predictive success is matched by its descriptive accuracy: that is, the extent to which it felicitously represents how attitudes come into being and how these attitudes causally interrelate with behaviour. When choices do not have to be made quickly, or are particularly important, or are irreversible, decision-making strategies may well be relatively analytic and akin to the weighing of probabilities and evaluations as implied by the processes of attitude formation as described within the theory of reasoned action. However, strategies may be qualitatively different when decisions are mundane, run-of-the-mill, in need of being made quickly and relatively inconsequential. It has to be recognised that such choices are often made using quick heuristics or 'rules-of-thumb' rather than a full-blown, analytic, compensatory weighing of different outcomes; however, one should not consider that these heuristics are generally unsuccessful, despite indications that they may be in certain circumstances (cf. Lopes, 1991).

We believe that the theory of reasoned action offers a useful model for the prediction of proximal psychological influences on food choice and that it is likely to benefit from some of the modifications highlighted above. That it requires to be complemented by integration with assessments of physiological, economic and social influences on food choice does not detract from its potential. Its predictive use has been clearly demonstrated. It also has tremendous practical potential through its identification of the psychological components (beliefs, values and attitudes) which need to be recognised and addressed in the application of dietary intervention practices. Moreover, the theory of reasoned action permits the assessment of people's attitudes to be based on an essentially rational, analytic model of human behaviour. 'Irrational' characterisations of people's choices imply a degree of unintelligibility: in the face of a plethora of commentaries concerning the irrationality of lay people's behaviour, such 'rational' models offer a more optimistic and challenging framework for empirical study.

References

Ajzen, I. (1988) *Attitudes, Personality, and Behavior*, Open University Press, Milton Keynes.

Ajzen, I. and Fishbein, M. (1980) *Understanding Attitudes and Predicting Social Behavior*, Prentice-Hall, Englewood Cliffs, New Jersey, pp. 70–1.

Ajzen, I. and Madden, T.J. (1986) Prediction of goal-directed behavior: Attitudes, intentions, and perceived behavioral control. *J. Exp. Soc. Psychol.*, **22**, 453–74.

Ajzen, I. and Timko, C. (1986) Correspondence between health attitudes and behavior. *Basic Appl. Soc. Psychol.*, **7**, 259–76.

Axelson, M.L., Brinberg, D., and Durand, J.H. (1983) Eating at a fast-food restaurant – A social-psychological analysis. *J. Nutr. Educ.*, **15**, 94–8.

Axelson, M.L., Federline, T.L., and Brinberg, D. (1985) A meta-analysis of food and nutrition-related research. *J. Nutr. Educ.*, **17**, 51–4.

Bargh, J.A. (1989) Conditional automaticity: Varieties of automatic influence in social perception and cognition, in J.S. Uleman and J.A. Bargh (eds) *Unintended Thought*, The Guildford Press, New York, pp. 3–51.

Beale, D.A. and Manstead, A.S.R. (1991) Predicting mothers' intentions to limit frequency of infants' sugar intake: testing the theory of planned behavior. *J. Appl. Soc. Psychol.*, **21**, 409–31.

Biddle, B.J., Bank, B.J., and Slavinge, R.L. (1987) Norms, preferences, identities and retention decisions. *Soc. Psychol. Quart.*, **50**, 322–37.

Brinberg, D. and Durand, J. (1983) Eating at fast-food restaurants: An analysis using two behavioral intention models. *J. Appl. Soc. Psychol.*, **13**, 459–72.

Carruth, B.R., Mangel, M., and Anderson, H.L. (1977) Assessing change-proneness and nutrition-related behaviors. *J. Am. Dietet. Assoc.*, **70**, 47–53.

Charng, H.-W., Piliavin, J.A., and Callero, P.L. (1988) Role identity and reasoned action in the prediction of repeated behavior. *Soc. Psychol. Quart.*, **51**, 303–17.

Committee on Medical Aspects of Food Policy (1984) *Diet and Cardiovascular Disease*, HMSO, London.

Cowart, B.J. and Beauchamp, G.K. (1986) Factors affecting acceptance of salt by human infants and children, in M.R. Kare and J.G. Brand (eds) *Interaction of the Chemical Senses with Nutrition*, Academic Press, Orlando, pp. 25–44.

Davidson, A.R. and Morrison, D.M. (1983) Predicting contraceptive behavior from attitudes: a comparison of within- versus across-subjects procedures. *J. Pers. Soc. Psychol.*, **45**, 997–1009.

Douglas, P.D. and Douglas, J.G. (1983) Nutritional knowledge and food practices of high school athletes. *J. Am. Dietet. Assoc.*, **84**, 1198–202.

Eagly, A.H. and Chaiken, S. (1993) *The Psychology of Attitudes*, Harcourt, Brace & Jovanovich, San Diego.

Eppright, E.S., Fox, H.M., Fryer, B.A., Lamkin, B.H., and Vivian, V.M. (1970) The North Central Regional study of diets of pre-school children. 2. Nutrition knowledge and attitudes of mothers. *J. Home Econ.*, **62**, 327–32.

Feldman, R.H.L. and Mayhew, P.C. (1984) Predicting nutrition behavior: The utilization of a social psychological model of health behavior. *Basic Appl. Soc. Psychol.*, **5**, 183–95.

Fishbein, M. and Ajzen, I. (1975) *Belief, Attitude, Intention and Behavior. An Introduction to Theory and Research*, Addison-Wesley, Reading, MA.

Fishbein, M. and Stasson, M. (1990) The role of desires, self-predictions, and perceived control in the prediction of training session attendance. *J. Appl. Soc. Psychol.*, **20**, 173–98.

Foley, C.S., Vaden, A.G., Newell, G.K., and Dayton, A.D. (1983) Establishing the need for nutrition education: III. Elementary students' nutrition knowledge, attitudes, and practices. *J. Am. Dietet. Assoc.*, **83**, 564–8.

Granberg, D. and Holmberg, S. (1990) The intention-behavior relationship among U.S. and Swedish voters. *Soc. Psychol. Quart.*, **53**, 44–54.

Grotkowski, M.L. and Sims, L.S. (1978) Nutritional knowledge, attitudes, and dietary practices of the elderly. *J. Am. Dietet. Assoc.*, **72**, 499–506.

Jalso, S.B., Burns, M.M., and Rivers, J.M. (1965) Nutritional beliefs and practices. *J. Am. Dietet. Assoc.*, **47**, 263–8.

James, W. (1983) *The Principles of Psychology*, Harvard University Press, Cambridge, Mass.

Kahle, L.R. and Beatty, S.E. (1987) The task situation and habit in the attitude-behavior relationship: A social adaptation view. *J. Soc. Behav. Pers.*, **2**, 219–32.

Khan, M.A. (1981) Evaluation of food selection patterns and preferences. *CRC Crit. Rev. Food Sci. Nutr.*, **15**, 129–53.

Lopes, L.L. (1991) The rhetoric of irrationality. *Theory and Psychol.*, **1**, 65–82.

McAlister, D.W., Mitchell, T.R., and Beach, L.R. (1979) The contingency model for the selection of decision strategies: An empirical test of the effects of significance, accountability, and reversibility. *Organizat. Behav. Hum. Perf.*, **24**, 228–44.

McEwan, J.A. and Thomson, D.M.H. (1988) An investigation of factors influencing consumer acceptance of chocolate confectionery using the repertory grid method, in D.M.H. Thomson (ed.) *Food Acceptability*, Elsevier Applied Science, London, pp. 347–61.

McGuire, W. (1985) Attitudes and attitude change, in G. Lindzey and E. Aronson (eds) *The Handbook of Social Psychology, Vol. 2. 3rd edition.*, Random House, New York, pp. 233–346.

Mittal, B. (1988) Achieving higher seat belt usage: The role of habit in bridging the attitude–behavior gap. *J. Appl. Soc. Psychol.*, **18**, 993–1016.

Perron, M. and Endres, J. (1985) Knowledge, attitudes, and dietary practices of female athletes. *J. Am. Dietet. Assoc.*, **85**, 573–6.

Pilgrim, F.J. (1957) The components of food acceptance and their measurement. *Am. J. Clin. Nutr.*, **5**, 171–5.

Randall, E. and Sanjur, D. (1981) Food preferences – Their conceptualization and relationship to consumption. *Ecol. Food Nutr.*, **11**, 151–61.

Rogers, P.J. and Blundell, J.E. (1990) Endogenous influences on food choices. *BNF Nutr. Bull.*, **15**, Suppl. 1, 31–40.

Rogers, P.J., Green, M., and Edwards, S. (1994) Nutritional influences on mood and cognitive performance: Their measurement and relevance to food acceptance (this volume).

Ronis, D.L., Yates, J.F., and Kirscht, J.P. (1989) Attitudes, decisions, and habits as determinants of repeated behavior, in A. Pratkanis, S. Breckler and A. Greenwald (eds) *Attitude Structure and Function*, Erlbaum, Hillsdale, NJ, pp. 213–39.

Rozin, P. (1986) One-trial acquired likes and dislikes in humans: Disgust as a US, food predominance, and negative learning predominance. *Learn. Motiv.*, **17**, 180–9.

Schifter, D.E. and Ajzen, I. (1985) Intention, perceived control, and weight loss: An application of the theory of planned behavior. *J. Pers. Soc. Psychol.*, **49**, 843–51.

Schwartz, N.E. (1975) Nutritional knowledge, attitudes, and practices of high school graduates. *J. Am. Dietet. Assoc.*, **66**, 28–31.

Shepherd, R. (1985) Dietary salt intake. *Nutr. Food Sci.*, **96**, 10–11.

Shepherd, R. (1987) The effects of nutritional beliefs and values on food acceptance, in J. Solms, D.A. Booth, R.M. Pangborn, and O. Raunhardt (eds) *Food Acceptance and Nutrition*, Academic Press, London, pp. 387–402.

Shepherd, R. (1988) Belief structure in relation to low-fat milk consumption. *J. Hum. Nutr. Dietet.*, **1**, 421–8.

Shepherd, R (1989) Factors influencing food preferences and choice, in R. Shepherd (ed.) *Handbook of the Psychophysiology of Human Eating*, Wiley, Chichester, pp. 3–24.

Shepherd, R. and Farleigh, C.A. (1986a) Attitudes and personality related to salt intake. *Appetite*, **7**, 343–54.

Shepherd, R. and Farleigh, C.A. (1986b) Preferences, attitudes and personality as determinants of salt intake. *Hum. Nutr. Appl. Nutr.*, **40A**, 195–208.

Shepherd, R. and Stockley, L. (1985) Fat consumption and attitudes towards food with a high fat content. *Hum. Nutr. Appl. Nutr.*, **39A**, 431–42.

Shepherd, R. and Stockley, L. (1987) Nutrition knowledge, attitudes, and fat consumption. *J. Amer. Dietet. Assoc.*, **87**, 615–9.

Sheppard, B.H., Hartwick, J., and Warshaw, P.R. (1988) The theory of reasoned action: A meta-analysis of past research with recommendations for modifications and future research. *J. Consum. Res.*, **15**, 325–43.

Sparks, P., Hedderley, D., and Shepherd, R. (1991) Expectancy-value models of attitudes: a note on the relationship between theory and methodology. *Europ. J. Soc. Psychol.*, **21**, 261–71.

Sparks, P., Hedderley, D., and Shepherd, R. (1992) An investigation into the relationship between perceived control, attitude variability and the consumption of two common foods. *Europ. J. Soc. Psychol.*, **22**, 55–71.

Sparks, P. and Shepherd, R. (1992) Self-identity and the theory of planned behavior: assessing the role of identification with 'green consumerism'. *Soc. Psychol. Q.*, **55**, 4, 388–99.

Tesser, A. and Shaffer, D.R. (1990) Attitudes and attitude change. *Ann. Rev. Psychol.*, **41**, 479–523.

Towler, G. and Shepherd, R. (1990) Development of a nutritional knowledge questionnaire. *J. Hum. Nutr. Dietet.*, **3**, 255–64.

Triandis, H.C. (1977) *Interpersonal Behavior*. Brooks/Cole, Monterey.

Triandis, H.C. (1980) Values, attitudes, and interpersonal behavior, in H.E. Howe and M.M. Page (eds) *Nebraska Symposium on Motivation*, 1979, University of Nebraska, Lincoln, pp. 195–259.

Tuorila, H. (1987) Selection of milks with varying fat contents and related overall liking, attitudes, norms and intentions. *Appetite*, **8**, 1–14.

Tuorila, H. and Pangborn, R.M. (1988) Behavioural models in the prediction of consumption of selected sweet, salty and fatty foods, in D.M.H. Thomson (ed.) *Food Acceptability*, Elsevier Applied Science, London, pp. 267–79.

Werblow, J.A., Fox, H.M., and Henneman, A. (1978) Nutrition knowledge, attitudes, and food patterns of women athletes. *J. Am. Dietet. Assoc.*, **73**, 242–5.

WHO (1991) *Diet, Nutrition, and the Prevention of Chronic Diseases*, WHO, Geneva.

Wicker, A.W. (1969) Attitude versus actions: The relationship of verbal and overt behavioral responses to attitude objects. *J. Soc. Issues*, **25**, 41–78.

Wittenbraker, J., Gibbs, B.L., and Kahle, L.R. (1983) Seat belt attitudes, habits, and behaviors: An adaptive amendment to the Fishbein model. *J. Appl. Soc. Psychol.*, **13**, 406–21.

Yudkin, J. (1956) Man's choice of food. *Lancet*, **i**, 645–9.

Zajonc, R.B. (1980) Feeling and thinking: Preferences need no inferences. *Am. Psychol.*, **35**, 151–75.

9 Nutritional influences on mood and cognitive performance: their measurement and relevance to food acceptance

P.J. ROGERS, M.W. GREEN and S. EDWARDS

9.1 Introduction

There is considerable public and scientific interest in the effects of diet on mood and behaviour. To date, however, work in this area has produced relatively few substantiated findings or definite conclusions. It is also unfortunate that in the public arena the results of scientifically valid studies have been largely overshadowed by extravagant and often emotive claims made in less objective work (e.g. see Lipton *et al.*, 1979 for a review of early work on food additives in relation to hyperactivity in children). There is little doubt that substances such as caffeine and alcohol have significant and important effects on mood and behaviour, and plausible hypotheses concerning the effects of other food components have been suggested, and in certain instances confirmed in properly conducted studies (Spring, 1986). There are a number of reasons why work on diet–behaviour relationships continues to provoke controversy. The purpose of the present chapter is to examine some of the prominent conceptual and methodological issues in this area, with particular reference to the effects of food and food constituents on mood and mental performance (cognitive efficiency). However, first it is important to consider the interrelationships between dietary effects on behaviour and food acceptance.

9.2 Relevance to food acceptance of dietary effects on behaviour

Figure 9.1 presents a simple scheme for conceptualising the relationships between diet, behaviour and food preference. Obviously, an important objective of work in this area is to identify effects of diet on behaviour. However, where such effects exist, presumably they must be mediated by physiological mechanisms. Therefore an understanding of diet–behaviour relationships requires that these mechanisms are specified. This makes the area of great scientific interest because it provides an important

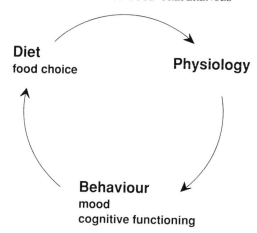

Figure 9.1 Interrelationships between diet and behaviour: dietary influences on mood and cognitive efficiency are mediated by physiological mechanisms and may feed back to influence food preference and consumption.

opportunity for examining the link between the biological and psychological domains of functioning.

The arrow going from 'behaviour' to 'diet' in Figure 9.1 indicates that the effects of food on mood and behaviour can influence food acceptance. This may occur through a conditioning process similar to that which has been investigated extensively in studies examining the development of learned aversions for foods producing sickness and the acquisition of preferences for foods conferring nutritional benefit. Such learned preferences and aversions appear to be characterised by alterations in the perceived pleasantness or liking of the foods' taste, flavour, texture, etc. (Rogers and Blundell, 1990). Whether or not sensory preferences can be conditioned by the effects of food on mood and cognitive performance is unknown, since the appropriate studies are not available (but see the section below on caffeine). Indeed, there are altogether very few data on preference conditioning in humans. Nonetheless, it appears likely that this is a mechanism by which, for example, the mood effects of food could have an important influence on food choice. Wurtman and Wurtman (1989) have argued that there is an increase in the consumption of high carbohydrate foods in seasonal affective disorder (SAD) and premenstrual syndrome. This change in nutritional behaviour is supposed to alter uptake of tryptophan into the brain and brain serotonergic activity, and in turn ameliorate the disturbances of mood associated with these conditions (see below for fuller details). The increase in carbohydrate intake appears to be driven by an increase in preference (liking) for high carbohydrate foods which has been termed 'carbohydrate craving'. These ideas are still controversial; for example, it is unclear whether the change in preference is for high carbohydrate or high fat foods or both. Nonetheless, they provide a good illustration of one possible functional relationship between the mood effects of food and food preference which, furthermore, is based on predictions about the physiological effects of food.

In addition to preference conditioning, conscious decisions will undoubtedly play an important role in determining food acceptance. The belief that a product has detrimental or undesirable effects may lead to its avoidance. Conversely, if an individual believes that consumption of a particular food or drink has beneficial effects, presumably they will be more willing to choose that product. Therefore, identifying nutritional effects on behaviour may be of considerable commercial value in product development and marketing. For example, 'Jolt Cola', which contains a high level of caffeine, is promoted on the basis of a claim that it can increase alertness and counteract fatigue. Although very few widely available products are currently promoted in this way, there is a rapidly growing interest in the concept of 'functional foods' and it seems likely that, if for no other reason than this, nutritional effects on mood and cognitive performance will receive increasing attention in the future.

9.3   Conceptual issues

9.3.1   Mood and emotion

It is surprising that, despite the large number of studies which have examined the effects of food and food constituents on mood, there has been almost no discussion in this literature of conceptual issues relating to mood and emotion, to problems of measurement, or to how changes in mood might affect thinking and behaviour. Frequently it appears that mood is viewed as being that which is measured by mood self-report scales (see below). One consequence has been that mood and arousal are confused – a change in alertness is often described as an effect on mood. However, there is an extensive literature on mood and emotion from the field of cognitive psychology. It is not possible to review this here, therefore the present discussion will focus on a recent theory of emotion which combines cognitive psychological and biological elements. Such a theory has the advantage that it can provide a basis for understanding how the physiological effects of food might influence psychological processes.

Traditional psychological investigations of 'emotion' concentrated on the perspective that emotion states are coincidental phenomena, which are irrelevant to the main body of a person's mental activity. Darwin (1872) concluded that emotions were accidental phenomena which result

Table 9.1 Five basic emotions together with their eliciting events and their effects

Emotion	Eliciting event	Effect
Euphoric		
Happiness	Sub-goal being achieved	Continue with plan, modifying as necessary
Dysphoric		
Sadness/depression	Failure of major plan or loss of active goal	Do nothing/search for new plan
Anxiety/fear	Self-preservation goal threatened	Stop, attend vigilantly to environment and/or escape
Anger	Active plan frustrated	Try harder, and/or aggress
Disgust	Gustatory goal violated	Reject substance and/or withdraw

From Oatley and Johnson-Laird, 1987.

from overflows of neural excitation and serve no function in the action of adult human beings. Recently, however, emotions have come to be regarded in terms of their functions within the human cognitive system. One example of this type of theoretical framework is the theory of Oatley and Johnson-Laird (1987) who postulate that there are five basic emotions (or 'emotion modes') which occur universally in humans. These correspond to happiness, sadness, anxiety, anger, and disgust (Table 9.1), and are viewed not merely as by-products of ongoing activity, but as an integral part of ongoing cognitive operations. The theory proposes that emotions are part of the biological solution to the problem of how to plan and carry out action aimed at satisfying multiple goals in environments which are not perfectly predictable. One function of emotions is to set the cognitive system suddenly into a particular mode and maintain it in that mode when an important goal is threatened. Thus fear or anxiety will be aroused when a self-preservation goal is threatened. For instance, while watching TV in an empty house you hear a door slam. You are frightened, and this activates a mode of preparedness for escape or other response and maintains a state of wary vigilance.

A second function attributed to emotions is that they occur when planned behaviour is interrupted (e.g. when a plan is thwarted) to facilitate the organisation of a new set of planned activities. Emotions also have a communicative role in the conduct of mutual plans, and it is argued that complex emotions (e.g. jealousy or remorse) are elaborations of the basic emotions within this social context.

Emotions are usually considered to be brief transitional phenomena, lasting for a few seconds or perhaps for a few minutes. This is based on evidence of physiological and facial changes accompanying emotions (Ekman, 1986). The phenomenology (experience) of emotion suggests, however, that emotional states may last longer than this, and it is here that it may be most appropriate to apply the term 'mood'. Oatley and Johnson-Laird (1987) refer to mood as a temporary predisposition to an emotion (in contrast to an enduring predisposition which would be related to 'temperament'), or an emotional state of low intensity capable of lasting for many minutes or hours.

9.3.2 Arousal, mood and cognitive performance

According to Schachter's (1964) theory of emotion, there are two components to any affective state, namely the state of arousal of the individual experiencing the emotion and the cognitions surrounding the emotion state. The cognitive component determines how the change in arousal level is interpreted, and there is little physiological differentiation between emotion states. This view has been extensively criticised, nonetheless it is clear that most affective states involve changes in arousal level. Other theoretical conceptualisations have also taken a similar dual component approach. An example is the two-component model of anxiety suggested by Liebert and Morris (1967) and Morris et al. (1981). They argue that the affective state of 'anxiety' has cognitive and physiological components which are termed 'worry' and 'emotionality', respectively. Unlike Schachter, though, they propose that these two factors are independent of each other and can be differentially aroused. A common feature of both these models is that a change in arousal level is a necessary component of an emotional experience, but this does not, on its own, constitute an emotional experience. An emotion consists of both arousal components and cognitive components, the latter being influenced by the individual's perception of the external events.

The close relationship between mood and arousal is also a central feature of Thayer's (1989) conceptualisation of mood, which proposes the existence of two arousal systems termed energetic and tense arousal (Table 9.2). For several reasons this approach appears especially relevant

Table 9.2 Thayer's (1989) model of mood and arousal: mood states are viewed as subjective components of two related biological arousal systems

Energetic arousal	Tense arousal
Sleep–wakefulness cycle	React to danger and threat
Mediates appetitive behaviour	Mediates escape and avoidance
Low – tiredness, fatigue	Low – calmness
High – energy, vigour	High – tension, anxiety (stress)
Positive hedonic tone increasing from low to high	Negative hedonic tone increasing from low to high

Interaction:
positively correlated at low to moderate levels
negatively correlated at high levels

to understanding nutritional and pharmacological effects on mood. In particular, it is concerned with short-term, non-pathological, fluctuations in mood and arousal, and their role in the regulation of adaptive behaviour. The energetic arousal system regulates the sleep–wakefulness cycle, and appears to be driven largely by endogenous factors according to a circadian rhythm, with peaks in energetic arousal occurring normally during the late morning and late afternoon. However, external events, physical exercise, food constituents and pharmacological agents can also strongly influence 'energetic moods'. The tense arousal system is activated when some real or imagined danger is present and functions to prepare the organism to make the appropriate responses. This includes preparation for 'fight or flight' as well as the regulation of restraint and inhibition.

Energetic and tense arousal are supposedly associated with different patterns of physiological and CNS activity, and therefore might be identified by, for example, biochemical or psychophysiological analysis. Most of the evidence for the existence of the two systems, however, comes from studies on self reports of mood-related bodily sensations and subjective feelings. Energetic arousal is recognisable by subjective feelings of energy, vigour and liveliness ('calm-energy'), and tense arousal by feelings of tension, anxiety and fearfulness ('tense-energy'). These and related feelings can be measured using the Activation–Deactivation Adjective Check List (Thayer, 1989). In its short form this consists of a list of 20 words (e.g. vigorous, drowsy, jittery, calm) which subjects rate on a 4-point scale.

An important aspect of energetic and tense arousal concerns the complex interaction of the two systems. It appears that a moderate amount of tension can raise feelings of energy, whereas high levels of tension can reduce energy. On the other hand, increasing energy can often have a tension-reducing effect. These interrelationships between energy and tension enable the theory to account for a variety of everyday feelings and occurrences. For instance, tension is often greatest at times of the day when energy is low; but is reduced when energetic arousal is increasing or at times of the day, such as late morning, when energy is high. Furthermore, Thayer (1989) argues that energetic and tense arousal are at the core of many other commonly identified moods. So that, for example, low energy and high tension are basic components of agitated depression, and high energy and low tension are associated closely with optimism, happiness and pleasurable bodily feelings. The latter might be identified simply as good mood, while tense-tiredness is unpleasant and might be identified as bad mood.

More relevant to the present discussion, Thayer suggests that caffeine may normally increase subjective energy indirectly by moderately increasing tension. However, at high doses or when tension is already high, negative effects may be induced. Such mixed effects of caffeine have been observed in a variety of studies (Thayer, 1989; James, 1991; and see below). On the other hand, an important action of alcohol may be to reduce tension. Another suggestion is that carbohydrate-snacking may be related to a desire to increase energetic arousal (Thayer, 1989), although this appears to contradict the hypothesis linking carbohydrate intake and brain serotonergic function (see above). The interaction between energetic and tense arousal provides a potentially powerful principle for explaining the use of foods and beverages to enhance mood. To date, however, only a very few of the studies on food and mood have adopted this particular theoretical perspective.

There can be little doubt that arousal level has an important effect on the way people perform on psychological tasks. One of the earliest, and most influential ideas on the exact nature of the relationship between task performance and arousal is known as the Yerkes–Dodson inverted-U hypothesis (1908), which is illustrated in Figure 9.2. Easterbrook (1959) proposed an explanation for this relationship which revolves around the idea of arousal causing a restriction in the amount of information which is gathered from the environment in the course of carrying out a task. According to Easterbrook, arousal acts to restrict the amount of outside information which is paid attention to in the completion of a task. Initially, this has a beneficial effect on task performance because a great

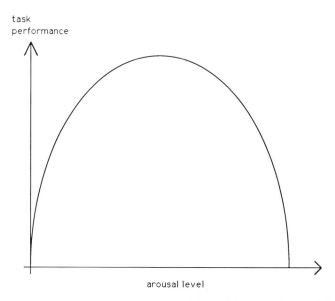

task
performance

arousal level

Figure 9.2 Proposed relationship between cognitive efficiency and arousal level (after Yerkes and Dodson, 1908).

deal of this environmental information is irrelevant to the task and detracts from optimum performance on that task. However, as arousal level increases further some information essential to the completion of the task is eliminated, and consequently, performance suffers.

A more recent account of the effects of arousal on task performance is that arousal acts to stimulate task-irrelevant, self-preoccupying processes in the subject. In the case of emotional arousal, this can be interpreted in terms of preferential processing for mood congruent material. Taking anxiety as an example, this means that anxious people should respond to evaluative testing conditions with ruminative self-worry and other negative task-irrelevant thoughts, a finding noted in numerous studies as far back as Mandler and Watson (1966).

9.3.3 Effects of food on mood and cognitive performance

How might food or food constituents influence mood and mental performance? The previous discussion suggests several possibilities. The sensory pleasure of eating could have important effects. The pleasurable or rewarding aspects of eating, perhaps mediated by endogenous opioid systems, might alter mood directly by producing a feeling of well-being (euphoria), or by distracting from or ameliorating dysphoric mood states such as anxiety and depression. In turn, relief from distressing mood states would be expected to reinforce preference for the food or drink consumed.

Food constituents may influence directly the activity of physiological and neurochemical systems subserving mood and emotional responses. A good example to illustrate such an interaction is the hypothesised relationship between carbohydrate preference, brain serotonergic activity and depressed mood. The idea that ingestion of carbohydrate might increase brain serotonin activity and thereby influence behaviour arose from work conducted in the early 1970s (Fernstrom and Wurtman, 1972). It was found that a carbohydrate meal can increase the plasma concentration of the amino acid tryptophan (TRP) relative to the other large neutral amino acids (LNAA). Insulin released in response to the carbohydrate load facilitates the uptake of most amino acids, but not tryptophan, into peripheral tissues such as muscle. Tryptophan is the precursor of the neurotransmitter serotonin. Since tryptophan and the other large neutral amino acids compete for entry into the brain and the rate-limiting enzyme for serotonin production (tryptophan hydroxylase) is not fully saturated with substrate under normal conditions, an increase in the plasma TRP/LNAA ratio can lead to an increase in brain serotonin. It was suggested that this increase in serotonin availability might increase serotonergic neurotransmission, leading to behavioural changes such as decreased vigilance and facilitated sleep onset. There is some evidence to support these proposed interrelationships, and also the possibility that eating carbohydrates can have positive effects in certain individuals suffering prolonged dysphoric mood states – where a functional deficiency of brain serotonergic activity might be implicated (Spring, 1990). However, it should be noted that the effects reported are often relatively weak and, furthermore, that certain assumptions about the neurochemical mechanisms involved remain largely speculative (Leathwood, 1987).

Another route by which food might influence mood and behaviour via a direct physiological effect is through the pre- or post-absorptive action of peptides derived from food proteins. Certain peptides have very potent biological effects, and it is apparent that, for example, peptides having opioid activity can be absorbed intact across the gastrointestinal tract (Gardener, 1988; Meisel and Schlimme, 1990). Recent studies have also demonstrated a potent effect on appetite of the dipeptide sweetener aspartame, which may be due to a direct effect on the release of the hormone cholecystokinin from the upper small intestine (Rogers *et al.*, 1991).

Any scientific study of diet and behaviour interactions should be guided by mechanistic hypotheses. However, different levels of analysis are possible, so that an explanation in terms of, for instance, a psychological mechanism or process can be just as useful as an explanation involving a neurochemical mechanism. Indeed, physiological and neurochemical hypotheses are often inappropriate, since there is almost no understanding of the physical basis of higher mental function. This point is illustrated by a recent attempt to understand the effect of alcohol on mood and behaviour. As an explanation of how alcohol makes social responses more extreme, enhances important self-evaluations, and relieves anxiety and depression, Steele and Josephs (1990) propose that alcohol produces a specific impairment of perception and thought which they call 'alcohol myopia'. It contrasts with other theories in this field which have emphasised expectations associated with the use of alcohol or the direct pharmacological effects of alcohol in causing specific reactions. The theory is successful in that it is able to account for what appeared previously to be conflicting findings, and because it makes clear predictions for further studies.

Finally, it is important to recognise that effects of food on mood and cognitive performance will be in many instances 'state-dependent'. For example, the effects of a carbohydrate meal may be to improve mood and performance, or to decrease vigilance depending on the individual (see above). The effect of caffeine may be to increase alertness (e.g. in the post-lunch period) and consequently improve performance on a task requiring sustained attention. However, caffeine may be found to have only a small or even a detrimental effect on performance if the same individual were tested when alertness (arousal) is already high – near optimum for that particular task. Similarly, the effects of food on mood are likely to be highly dependent on the subject's basal mood state. The anxiety-reducing effect of an anxiolytic substance will be strongest under conditions producing high levels of anxiety. The implication of this state-dependency is that experimental studies must at the very least measure and control baseline mood and/or level of arousal. Both the quality and intensity of mood should be identified. Almost certainly the most effective method would be to use a mood manipulation technique (see below), which together with appropriate experimental designs can be used to test, for example, the anxiolytic or anti-depressant effects of consuming a particular food or food constituent. To date there are almost no studies in this field that have used such an approach (but see Young, 1986).

9.4   Methodological issues

9.4.1   Experimental design

The most straightforward and powerful experimental design that can be used to investigate the effects of food constituents on mood and performance is the double-blind, cross-over design. The food constituent of interest might be administered orally in a capsule and its effects compared with that of an inert placebo of identical appearance. Given the large inter-subject variability in most psychometric and psychophysiological responses, a within-subjects (repeated measures) design with $n = 12-20$ is an efficient design (Kraemer, 1981). The order of treatments is, of course, counterbalanced across subjects. This design minimises effects of the subjects' (and experimenters') expectations, and provides the opportunity to manipulate expectations independently if required. Such an approach has proved successful in work examining the post-ingestive effects of intense sweeteners (Rogers *et al.*, 1991).

If an experimental manipulation requires the administration of relatively large amounts of a food constituent it is not practical or desirable to use the above method. For instance, large doses of up to 10 g of amino acids have been administered in this way (Ryan-Harshman *et al.*, 1987), but subjects were required to consume 24 gelatin capsules and 500 ml water to wash them down! A more satisfactory method is to use a food or drink as the vehicle for the manipulation. For example, tryptophan has been administered successfully using a tryptophan supplemented chocolate bar which was indistinguishable organoleptically from the placebo bar (Hill and Blundell, 1988). Carbohydrate loads can be disguised by using bland, highly soluble starches such as low dextrose equivalent maltodextrins in a food base, by matching the sweet taste of a sugar supplemented food or drink using an intense sweetener, or by adding a sweetness inhibitor such as lactisole (Rogers and Blundell, 1989 and unpublished; Saravis *et al.*, 1990).

It is important to recognise that manipulations using capsules and a food or drink base are not equivalent, even where the appearance, taste, etc. of active and control treatments are well matched. Thus a food vehicle contains nutrients which could interact with the substance under investigation. For example, the effect of carbohydrate ingestion on plasma tryptophan to large neutral amino acid ratio (see above) is markedly reduced if only a small amount of protein is included in the meal (Teff *et al.*, 1989). Furthermore, consumption of a food or drink will evoke certain expectations and learned responses which are either eliminated or are different when capsules are administered. Thus, irrespective of possible differences in the rate of absorption from the gastro-intestinal tract, does caffeine given in a capsule (versus placebo capsule) have the same effects as caffeine given in coffee (versus decaffeinated coffee)?

In the designs described above it is possible to control and/or manipulate expectations and learned responses because subjects are blind to the experimental manipulation. There is no such control in an open design where, for example, the effects of consuming a food are compared against a baseline, or a second experimental condition in which no food or an organoleptically different food is given. Accordingly, only very general conclusions can be drawn from the results of open trials. For example, although the findings of open clinical trials on the effects of placing hyperactive children on the Feingold diet were regarded as supporting the detrimental effects of certain food additives, subsequent double-blind challenge studies failed to confirm this conclusion. Changing the diet had a clinically significant effect on hyperactivity, but probably this was due largely to expectancies and changes in family interactions rather than to a specific action of food constituents (Lipton *et al.*, 1979).

9.4.2   Sensitivity and validity of measures

A methodological problem which is important, yet often overlooked, concerns the sensitivity of measures of mood, performance and behaviour. The detection of a statistically significant treatment effect on a particular measure allows a positive conclusion to be drawn (e.g. caffeine improves sustained attention in the post-lunch period). The failure to detect a treatment effect, however, does not provide unequivocal evidence that the treatment is benign. The measures used might simply have been insensitive or misapplied. A 'no effect' conclusion can be strengthened in a number of ways. For example, using measures of known sensitivity (i.e. measures which have proved sensitive to other treatments), or by including a 'positive' control treatment within every study or at the outset of a series of related studies. An example of the latter would be the use of caffeine as a third treatment in a study comparing the effects of placebo and a substance with suspected stimulant properties. This approach is the most powerful. Furthermore, it contributes to establishing the practical significance of a positive outcome, which depends on the size of the effect observed and also the validity of the measures employed. Using the same example, if the new substance is found to have low potency compared with caffeine and/or the experimental tests employed have low validity (e.g. they are inadequate measures of the cognitive processes under investigation, or they do not relate clearly to real-life tasks) then the importance of any treatment effects is considerably diminished.

9.4.3 Mood measures

There are two main types of measure which can be used to determine the effects of food and food constituents on mood and arousal level, namely questionnaire based measures, and performance based measures. The relative merits of each of these two approaches are discussed below.

Research on the effects of different food constituents on mood has to date relied almost exclusively on questionnaire based measures, whether these be somewhat *ad hoc* measures or experimentally validated mood scales such as the Profile of Mood States Questionnaire (POMS) of McNair *et al.* (1971). POMS is a mood questionnaire consisting of 65 adjectives, each rated by test subjects on a five-point scale. Factor analysis yields six factors: tension–anxiety, depression–dejection, anger–hostility, vigour–activity, fatigue–inertia and confusion–bewilderment. Perhaps the simplest self-report measure for assessing mood is the Visual Analogue Mood Scale (VAMS) of Folstein and Luria (1973). The VAMS comprises a rectangular card, 100 mm by 35 mm, on which the following instruction is printed: 'How is your mood right now? A mark on the line toward the left represents your worst mood, toward the right, your best.' VAMS is scored by measuring the distance from the left end of the card to the subject's mark. This measure has been used in many studies on the effects of dietary manipulation on mood and performance (e.g. Lieberman *et al.*, 1983; Spring *et al.*, 1983). A more detailed questionnaire based on visual analogue format was developed by Bond and Lader (1974; and see Smith *et al.*, 1991b). This consists of a series of bi-polar visual analogue scales including, drowsy–alert, tense–calm, depressed–elated and interested–bored.

Even though these have to some extent been validated, such self-report scales suffer from several weaknesses. For example, it is difficult to conceal from subjects that mood changes are under investigation. Exposure to the questionnaire material may itself influence mood and/or produce expectations which will then seriously bias the subjects' responses. (Although a suitable cover story can be used, in which subjects are told, for example, that the purpose of the questionnaire is to 'control for differences in mood', this is at best only a partial solution to the problem.) Related to this, there are inherent problems with any type of questionnaire measure which does not take account of impression management, that is the tendency of subjects to tailor their questionnaire responses in order to present themselves in the best possible light or to alter their responses according to how they believe the experimenter wants them to respond. An example of this effect can be seen in the study of Power and MacRae (1977), who found that individuals could identify the items necessary to manipulate their responses on the Eysenck Personality Inventory in order to present any personality profile that they were instructed to. In the case of self report measures relating to mood or emotion, there is evidence that impression management is a factor which has an effect on subjects' responses. Flett *et al.* (1988) correlated subjects' self reports of emotional experiences with measures of impression management and self-deception. It was found that the impression management and self-deception measures were significantly correlated with subjects' reports of the frequency, intensity and duration of negative emotions. In the case of mood questionnaires such as the VAMS or POMS, there are obvious problems with the social desirability to subjects of presenting themselves as being anxious or depressed.

With one notable exception (Young, 1986), performance based measures of mood do not appear to have been used in the research examining the effects of food and food constituents. Young (1986) describes a study in which subjects were fed either a balanced amino acid mixture, or amino acid mixtures that were either tryptophan-free or tryptophan-supplemented. The mood measure was based on distractibility during a proofreading task. The subjects performed this task while listening to tapes of varying emotional content over headphones. They were tested under no distraction, low distraction (readings from a statistics textbook), high distraction (eyewitness accounts of the bombing of Hiroshima) and 'dysphoric' distraction (themes of hopelessness and helplessness). In the tryptophan-depleted group proofreading performance was significantly worse with the dysphoric distracter than with the low distracter. On the assumption that individuals who are depressed will be more distracted by dysphoric themes than people who are not depressed, this result supports the prediction that the tryptophan-free mixture would depress mood, which was further confirmed in self-report measures.

A number of other performance based measures of mood have been developed in work unrelated to the effects of food. For example, in the semantic judgement task, subjects are presented with a series of sentences on a VDU, each of these sentences having a word-long gap in it. While the sentence is still on the screen, a single word appears on the screen below the sentence. The task is to decide whether, when the single word is inserted into the gap in the sentence, the sentence is then semantically and grammatically correct. Subjects are provided with an option of two buttons to press, one marked 'SENSIBLE' and one marked 'NON-SENSIBLE'. They have to press the appropriate button for each sentence as quickly as possible, and the dependent variable is the time taken to do this. The rationale behind this task is that, because there is an attentional bias towards the encoding of mood congruent material, subjects should be faster to respond to those stimulus words which are congruent with their induced mood state. If the test substance has the effect of altering mood, for example an anxiety reducing effect, then this should be revealed in the form of a reduced difference between the performance on mood congruent and mood incongruent words for subjects administered the test substance compared with subjects receiving placebo, who would be expected to demonstrate a strong mood congruent bias.

In the dot-probe task subjects are presented with a series of pairs of words on a VDU. Subjects are required to name aloud the uppermost word of the pair for each pair. They also have a secondary task of pressing a button as fast as possible whenever they see a dot, which appears at random intervals in place of the lowermost word of the pair. It has been found (MacLeod et al., 1986) that anxious subjects are faster to respond to this dot when it replaces a threat word than a non-threat word. The interpretation of this result is that attention is automatically drawn to the processing of mood congruent material. In the case of an anxiety reducing substance, subjects receiving this substance should show a smaller effect than those receiving placebo.

The 'emotional' Stroop task is a paradigm in which subjects have to name the colours of a series of words presented to them. Subjects take longer to colour-name mood congruent words than a set of mood incongruent, neutral control words (e.g. Watts et al., 1986). As with the previous two tasks, this is a finding consistent with the idea of allocation of attentional resources to the processing of mood congruent material. Because attentional resources are removed from the task which has been set (i.e. colour-naming) in order to process the semantic content of the mood congruent word, colour-naming performance with these words is poorer (slower) than with mood incongruent words. The paradigm can therefore be used as an index of mood state. It is worth noting that an anxiety-reducing substance should promote faster responding to mood congruent material in this third paradigm, but slower responding in the first two paradigms. Therefore, used together, these methods provide a way of controlling for non-specific effects on performance.

Performance based procedures provide a potentially powerful method for assessing mood changes, and avoid many of the limitations associated with self-report measures. However, these tasks necessarily take longer to complete, and would generally be less suitable in screening for unspecified mood changes or where there is no experimental control over mood. The main difficulty is to include stimulus materials which are congruent with a large enough number of affective states. Some preliminary work has shown that the 'emotional' Stroop task is perhaps the most adaptable of the various performance based measures and consequently this task may be developed as a valuable tool in the assessment of nutritional effects on mood (Green and Rogers, 1992).

9.4.4   Mood induction

It was argued earlier that a weakness of many studies examining dietary effects on mood has been the lack of control over mood. One approach to this problem is to use a mood induction technique, some examples of which are discussed briefly here.

One of the earliest, and most frequently used techniques for inducing affective states is the Velten induction procedure (VIP) (Velten, 1968). In this procedure, the subjects read a series of cards, each containing a self-relevant statement calculated to induce the mood required. An example of this would be the statement 'I FEEL DOWNHEARTED TODAY', which would be expected to induce a depressed mood. There are, however, several important problems with the VIP; for example, there is a large inter-individual variability in response to the procedure. Polivy and Doyle (1980) found that only 50% of their subject population reported actually feeling the moods implied by the statements with which they were presented. Using a within-subjects design, Sirota and Schwartz (1982) found that only 70% of their subject population met the mood change criterion set, while Teasdale and Taylor (1981), using slightly stricter criteria, found that only 54% of their subjects showed a difference in self-reported mood state after the VIP.

One practical implication of these findings is that large numbers of subjects need to be tested in order to obtain a reasonable pool of people who demonstrate effects on the VIP, and this raises questions about the extent to which the results obtained using the VIP can be generalised. Thus subjects who do respond to the VIP are not a random sample of the population. Scheier and Carver (1977) found that subjects who scored high on Fenigstein et al.'s (1975) private self-consciousness scale, responded more strongly to the VIP depression induction than low scorers. It is possible that individuals who score highly on this scale are more aware of their moods and feelings, and are generally more self-reflective and introspective. The VIP has also been criticised for being prone to demand characteristics. Buchwald et al. (1981) have suggested that the effects observed with the VIP may be artefactual because subjects may only report a change in mood state in order to comply with the demands inherent in the experimental situation.

There are, however, other means of experimentally inducing mood states which appear to have greater face validity than the VIP. For instance, Clark (1983) has reported that appropriate music played to subjects, prior to testing, is an effective method of inducing mild mood states, and Morris et al. (1981) have found that showing subjects video-tapes of peer groups exhibiting the appropriate type of behaviour for a certain mood is an effective way of inducing that mood. A number of methods for arousing situationally specific anxiety in subjects have been developed, including threat of electric shock! (Herman and Polivy, 1975). Using procedures which are perhaps more acceptable ethically, Martens (1969) found that the presence of observers in the same testing room as subjects significantly elevated their anxiety ratings, and Meunier and Rule (1967) found that manipulating the type of feedback (positive or negative) on success in carrying out a task influenced the number and frequency of anxious, self-deprecatory thoughts reported by subjects. Another effective means of inducing anxiety is to make subjects believe that the task which they are about to complete is in some way a measure of intelligence (Sarason and Stoops, 1978; Larsen and Ketelaar, 1989).

9.4.5 Measures of cognitive performance

There is a fairly extensive literature on the effects of food and food constituents on what can be loosely termed 'cognitive' performance. At least some of this work has its origins in the interest in cognitive skill which arose in the 1950s and 1960s which resulted from a need to develop and improve weapon and radar systems following the Second World War. An essential aspect of the design of these was to understand the physical and mental capacities of the individuals operating them, and the best way to train the operators (Colley and Beech, 1989). Thus laboratory tasks were developed, for example, to simulate a weapons operator's ability to track a target or a radar operator's ability to track an aeroplane on an oscilloscope, and one area of particular interest was diurnal variations in performance. In relation to dietary influences, this led to studies examining the effects of eating lunch on the so called 'post-lunch dip' in performance efficiency. It is now clear that the macronutrient composition and the size of the meal can influence the post-lunch dip, although there are 'post-lunch' impairments on some tasks whether or not lunch is eaten (Craig, 1986; Smith et al., 1988, 1991a; and see below). At the same time the development of information processing models of human cognition led to a rapid growth in experimental studies of perception, memory, learning, attention and reasoning. In turn, many of the procedures that have been developed in this work have been adapted for use in determining the effects of, for example, dietary and pharmacological manipulations (e.g. Warburton and Rusted, 1991). A common method is to administer a battery of tasks to subjects with the intention of testing various aspects of cognitive performance. This literature will not be reviewed in detail here (see Craig, 1986; Spring, 1986; Smith, 1988; Warburton and Rusted, 1989); however, the examples given below indicate the scope and some of the limitations of this approach.

Perhaps the simplest performance measure is reaction time. Lieberman *et al.* (1986) used an auditory reaction time task in which subjects were required to respond as quickly as possible to the onset of a tone by lifting their finger from a key on a microcomputer keyboard. There were 125 trials, and the start of each trial was cued visually on the computer VDU. Reaction time was found to be significantly slower after a high carbohydrate lunch compared with an equicaloric high protein lunch. In another study this measure was found to be unaffected by caffeine, although caffeine administration significantly improved choice reaction time (Lieberman *et al.*, 1987). Choice reaction time is the response latency to any one of two or more distinct signals that may occur, where each signal requires different response (usually pressing a different key in an array paralleling an array of stimulus lights). Choice reaction time is greater than simple reaction time, and the time difference between these reaction times (i.e. mean choice reaction time minus mean simple reaction time) can provide a measure of the time required for the mental processes of discrimination and choice decision. This and the finding that reaction time increases as a linear function of the logarithm of the number of choice alternatives has been investigated extensively (Jensen, 1987), although these tasks have apparently not been used very widely in nutritional studies.

Attentional tasks are frequently included in cognitive performance test batteries. Sustained attention or vigilance can be assessed using tests requiring the detection of relatively rare events during a prolonged test session. For example, in a study examining the effects of caffeine on performance, subjects were given the task of detecting 40 randomly-timed signal tones embedded in 1800 slightly longer background tones (Lieberman *et al.*, 1987). The test lasted for 1 hour. Caffeine in doses ranging from as low as 32 mg to 256 mg substantially improved performance assessed by the number of targets detected. This task has high face validity as a measure of sustained attention: it is a simple task but requires the subject to attend closely and continuously. Results with a similar vigilance task (Detection of Repeated Numbers) using visually presented stimuli showed decrements in performance during the early afternoon which was due at least in part to eating lunch (Smith and Miles, 1986). A disadvantage is that these tests are very time consuming and fatiguing, which may preclude their inclusion in many studies. As a result other more economical procedures have been developed. These are rapid information processing tasks such as the Continuous Performance Test which consists of a series of digits or letters presented to the subject on paper (self-paced) or a computer VDU (experimenter-paced). The subject is required to respond to a designated number or letter occurring randomly in the list (Buchsbaum and Sostek, 1980). In more complex variants of this task, such as the Bakan task, the target is a sequence of more than one number or letter. For instance, Smith and colleagues (Smith and Miles, 1986; Smith *et al.*, 1990) used a task in which subjects were presented with a stream of single digits via a VDU at the rate of 100 digits per minute. Subjects were required to press a response button whenever they detected a sequence of three odd or three even digits. During the 10-minute test, eight such targets were presented every minute separated by a minimum of five and a maximum of 30 digits. Performance on this task (percent targets detected) was impaired after lunch compared with the pre-meal baseline; however the post-lunch dip in performance was completely removed by caffeine (3 mg/kg body weight). Although rapid information processing tasks have been shown to be sensitive to nutritional and also pharmacological manipulations (e.g. Warburton and Rusted, 1991), the interpretation of such results is perhaps less clear than that for simpler vigilance tasks. Thus the rapid information processing task used in the above studies would appear to have a large memory component (Smith, 1989). It may be significant therefore that effects of caffeine were not detected on the short duration, but low memory loaded Continuous Performance Test (Lieberman *et al.*, 1987).

Other attentional tasks are the Stroop task and the dot-probe task, versions of which have been described in the section on mood measures. In Stroop's (1935) original studies subjects were required to name the colours of a number of stimuli while attempting to ignore the stimulus itself. He found that it takes longer to name the colours when the base items are in the form of antagonistic colour names (e.g. 'BLUE' written in red ink) than when they are in the form of colour patches. Thus the semantic content of the base items (task-irrelevant information) interferes with the colour-naming task. Despite a great deal of subsequent work, this interference effect is by no means fully understood (MacLeod, 1991). Nonetheless, the Stroop task can perhaps be regarded as a measure of selective attention (Smith, 1989). Smith and Miles found that lunch did not influence the amount of interference found in the colour Stroop test. However, in another study, caffeine impaired performance (increased the difference between control and interference conditions) on a numerical version of the Stroop task (Foreman *et al.*, 1989). Spring *et al.* (1983) compared the effects of consuming carbohydrate-rich and protein-rich meals on selective attention, assessed using a dichotic listening task. In dichotic listening subjects wear stereo headphones and repeat back immediately (shadow) the words presented to one ear. These words are sometimes presented alone and at other times they occur simultaneously with words presented to the other ear. Subjects who ate the carbohydrate-rich lunch shadowed less accurately, although this impairment occurred regardless of whether distraction was present or absent suggesting a decrement in sustained attention rather than selective attention. However, significant effects of meal composition on different aspects of selective attention have been reported in a subsequent study (Smith *et al.*, 1988). High starch and high sugar meals were found to slow reactions to peripheral visual stimuli, whereas for another task consumption of a high protein lunch was associated with greater susceptibility to distraction from stimuli close to the target.

Memory assessment has been reviewed recently by Warburton and Rusted (1989). Although this review takes a pharmacological perspective it is also highly relevant to nutritional studies and the material will not be repeated here. Using different information processing tasks along with appropriate study designs it is possible to assess various aspects of memory and memory processes, including verbal and spatial memory, the encoding, storage and retrieval of information, and state-dependent effects. However, despite the very considerable interest in memory research and, for example, concern about the deterioration of memory in old age and memory impairment in dementia, there has been little work attempting to identify effects of nutritional manipulations on memory (but see Smith, 1988; Kanarek and Swinney, 1989). This may therefore provide an important opportunity for future studies, although it is important to note that there appears to be no specific hypothesis predicting nutritional influences on memory.

This highlights several important issues concerning the approach and interpretation of studies examining performance (and mood) effects. Merely selecting a battery of measures and testing the effects of a nutritional manipulation can reveal relatively little. In the absence of specific hypotheses to determine the choice of tasks, it will, in general, be possible to conclude only that a particular manipulation produces a particular effect. As well as being restricted such a conclusion can mislead because the result obtained may be due to a change, for example in alertness, which has consequences for a number of aspects of performance. As Warburton and Rusted (1989) have noted the 'selection of experimental paradigms (tasks) should always be guided by their potential for distinguishing between theoretical alternatives' (p. 166). Indeed, finding a particular change in performance does not necessarily imply that a distinct cognitive process has been affected. Effects on performance can be due either to intrinsic factors which are related directly to cognitive operations, or to extrinsic factors which cannot be regarded as specific cognitive processes but which nonetheless influence cognition and produce behavioural change (Warburton and Rusted, 1989). This is a common problem in psychobiology, which can be addressed by the use of appropriate methodology. One approach is to compare several nutritional manipulations on a number of different tasks. The strongest result is where treatment 1 is found to affect ability A but not ability B, while treatment 2 affects B but not A. This is known as a double dissociation and indicates that the treatments are exerting specific effects (Weiskrantz, 1968). Performance test batteries can therefore provide a powerful tool for determining diet–behaviour relationships, however it is important that the choice of tasks is driven by theory rather than convenience.

9.5 Caffeine: a case study

There has been extensive research on the effects of caffeine on mood and cognitive performance, some of which has been discussed above. Here this work is examined in more detail because it provides a particularly good example of many of the conceptual and methodological issues involved in the study of diet–behaviour relationships. As a common constituent of beverages, foods and medications, there is a widespread interest in caffeine. Furthermore, it is possible to relate the effects of caffeine to specific physiological systems (see Milon et al., 1988, for review). At very high doses, caffeine is thought to stimulate muscle contraction by the mobilisation of intracellular calcium, and to facilitate post-synaptic adrenergic transmission via the inhibition of phosphodiesterase and elevation of cyclic adenosine monophosphate activity. However, these effects are not thought to occur at normal dietary levels of intake, at which caffeine acts as an adenosine receptor antagonist.

Although there is a large literature on the effects of caffeine on mood, there is a remarkable lack of agreement as to the findings. This may be due in part to the fact that much of this work is concerned with non-normal subject populations (e.g. 'caffeinism' sufferers), but also to the lack of control, in many studies, of variables thought to be important to the effects of caffeine (this point is discussed more fully below). The most generally reported 'mood' effects of caffeine are the generation of anxiety and arousal (e.g. Goldstein et al., 1965; Gilliland and Andress, 1981; Leathwood and Pollet, 1983; Loke, 1988; Stern et al., 1989). The anxiogenic effect of caffeine, however, has not always been found in normal populations (e.g. Johnson et al., 1990a,b). It may be significant, therefore, that caffeine pretreatment has been reported to increase anxiety ratings in a high- but not a low-stress situation (Shanahan and Hughes, 1986), because this suggests that caffeine may act to amplify existing anxiety rather than as a direct anxiogenic agent. Furthermore, as discussed above, arousal is only one of the components of mood. Johnson *et al.* (1990a,b) administered a benzodiazepine (either triazolam or flurazepam) or placebo at night and caffeine or placebo in the morning. Although the benzodiazepine and caffeine treatments both had clear effects on arousal, neither had any effect on self-reported mood (POMS or VAMS scores).

The interpretation of findings concerning the effects of caffeine on cognitive performance are also somewhat unclear. There is evidence to suggest that caffeine may ameliorate the depressive effects on performance of the post-lunch dip in arousal (Smith *et al.*, 1990; and see above), and that the effects of caffeine are dependent on the actual nature of the task under investigation (e.g. Foreman *et al.*, 1989). However, two new lines of work have recently produced some provocative data. First, it has been reported that caffeine, when administered in a capsule, can be distinguished from placebo at doses as low as 10 mg (Griffiths *et al.*, 1990). This contrasts with the much larger doses (typically greater than 200 mg) administered in most, though not all (e.g. Lieberman *et al.*, 1987) performance studies. Second, there is now an emerging body of evidence to suggest that caffeine, at doses such as those found in coffee, may act as a reinforcing stimulus in humans, but higher doses may act as an aversive stimulus (Griffiths *et al.*, 1986, 1989; Griffiths and Woodson, 1988). Thus, humans will self-select capsules or beverages at normal doses in preference to placebo capsules or decaffeinated beverages.

Why, then, has the immense amount of research designed to investigate the effects of caffeine failed to produce a convincing consensus on its influence on mood and cognitive performance? One problem concerns weaknesses in the conceptualisation and measurement of mood (see above). However, this does not account fully for the frequent, but by no means invariable, effects that have been reported. A further important problem, therefore, appears to be that most studies have failed to take account of the many variables which have been reported to influence the effects of caffeine. For example, the dosage of caffeine used seems to be important for a number of reasons. Of course, larger effects are expected at higher doses. However, the dose-response relationship is complex, since, as described above, very high doses may activate different physiological systems to low doses, producing qualitatively as well as quantitatively different psychological effects. Moreover, the dose of caffeine administered must be set in the context of habitual caffeine intake, since caffeine abstainers and high caffeine consumers may react quite differently to the same dose. Unfortunately, the assessment of habitual caffeine intake is difficult. Not only is caffeine frequently ingested unknowingly (e.g. in soft drinks not usually associated with caffeine, such as some brands of carbonated orange, or in medications containing aspirin), but even such obvious sources of caffeine as coffee are often difficult to evaluate for caffeine content, since this can depend on blend, year, date of production, roasting, and brewing habits, both individual and national (Milon *et al.*, 1988). In evaluating cognitive performance, there may be an interaction between task and dose, while differences related to personality traits (e.g. Smith *et al.*, 1991b) provide a yet further complication. Also the time of day of testing would appear to be crucial, because any effect of caffeine on arousal must be set in the context of the normal circadian rhythm of arousal.

It is clear that further work is required to evaluate fully the behavioural effects of caffeine. In particular the use of relatively low doses of caffeine would produce effects more comparable to everyday life than the administration of high bolus doses. Not only do humans take their caffeine in low doses normally, but discrimination studies suggest strongly that it is simply not necessary to use high doses to produce subjectively noticeable effects, and that very high doses may produce physiological effects which are irrelevant to normal consumption. Another priority is the investigation of the reinforcing effects of caffeine, since this has important implications for the role of caffeine in the acceptance of caffeine-containing foods and beverages (see section 9.2).

9.6 Summary

Despite considerable interest in research on the effects of food and food constituents on mood and mental performance there are a number of important conceptual and methodological problems which undermine much of the work carried out in this field. The present review examines the experimental assessment of mood and cognitive performance and suggests how research strategies may be improved. In particular it is argued that studies examining diet–behaviour relationships should be driven by specific, mechanistic hypotheses. Some recent developments in the measurement of mood are discussed, and it is suggested that these methods together with mood manipulation techniques may provide an important opportunity for improving the understanding of nutritional effects on mood. Finally, an issue that has been almost completely neglected, is the extent to which the behavioural effects of food influence food preference and food choice. Given its considerable practical and theoretical significance, it is suggested that this is another area which should be given priority in future research.

References

Anderson, K.J. and Reville, W. (1983) The interactive effects of caffeine, impulsivity and task demands on a visual search task. *Personality and Individual Differences*, **4**, 127–34.

Bond, A. and Lader, M. (1974) The use of analogue scales in rating subjective feelings. *Brit. J. Medical Psychol.*, **47**, 211–18.

Buchsbaum, M.S. and Sostek, A.J. (1980) An adaptive-rate continuous performance test: vigilance characteristics and reliability for 400 male students. *Perceptual and Motor Skills*, **51**, 707–13.

Buchwald, A.M., Stracks, S., and Coyne, J.C. (1981) Demand characteristics and the Velten mood induction procedure. *J. Consulting and Clinical Psychol.*, **49**, 478–9.

Bursill, A.E. (1958) The restriction of peripheral vision during exposure to hot and humid conditions. *Quarterly Journal of Experimental Psychol.*, **10**, 113–29.

Clark, D.M. (1983) On the induction of depressed mood in the laboratory: Evaluation and comparison of the Velten and musical procedures. *Advances in Behaviour Research and Therapy*, **5**, 24–49.

Colley, A.M. and Beech, J.R. (1989) Acquiring and performing cognitive skills, in A.M. Colley and J.R. Beech (eds) *Acquisition and Performance of Cognitive Skills*, Wiley, Chichester, pp. 1–16.

Craig, A. (1986) Acute effects of meals on perceptual and cognitive efficiency. *Nutrition Reviews*, **44** (Supplement), 1163–71.

Darwin, C. (1872) *The Expression of the Emotions in Man and the Animals*, reprinted, University of Chicago Press, Chicago.

Deffenbacher, J.L. (1978) Worry, emotionality, and task generated interference in test-anxiety: An empirical test of attentional theory. *Journal of Educational Psychology*, **70**, 248–54.

Easterbrook, J.A. (1959) The effect of emotion on cue utilisation and the organisation of behaviour. *Psychological Review*, **66**, 183–200.

Ekman, P. (1986) Expression and the nature of emotion, in K.P. Scherer and P. Ekman (eds) *Approaches to Emotion*, Lawrence Erlbaum, Hillsdale, NJ, pp. 319–44.

Fenigstein, A., Scheier, M.F., and Buss, A.H. (1975) Public and private self consciousness: assessment and theory. *Journal of Consulting and Clinical Psychology*, **43**, 522–7.

Fernstrom, J.D. and Wurtman, R.J. (1972) Brain serotonin content: physiological regulation by plasma neutral amino acids. *Science*, **178**, 414–41.

Flett, G.L., Blankstein, K.R., Pliner, P., and Bator, C. (1988) Impression-management and self-deception components of appraised emotional experiences. *British Journal of Social Psychology*, **27**, 67–77.

Folstein, M.F. and Luria, R. (1973) Reliability, validity and clinical application of the Visual Analogue Mood Scale. *Psychological Medicine*, **3**, 479–86.

Foreman, N., Barraclough, S., Moore, C., Mehta, A., and Madon, M. (1989) High doses of caffeine impair performance of the numerical stroop task in men. *Pharmacology Biochemistry and Behaviour*, **32**, 399–403.

Gardener, M.L.G. (1988) Intestinal absorption of peptides, in J.E. Morley, M.B. Sternman, and J.H. Walsh (eds) *Nutritional Modulation of Neural Function*, Academic Press, San Diego, pp. 29–38.

Geller, V. and Shaver, P. (1976) Cognitive consequences of self-awareness. *Journal of Experimental Social Psychology*, **12**, 99–108.

Gilliland, K. and Andress, D. (1981) Ad lib caffeine consumption, symptoms of caffeinism, and academic performance. *American Journal of Psychiatry*, **138**, 512–14.

Goldstein, A., Kaizer, S., and Warren, R. (1965) Psychotropic effects of caffeine in man. II. Alertness, psychomotor coordination, and mood. *Journal of Pharmacology and Experimental Therapeutics*, **150**, 146–51.

Green, M.W. and Rogers, P.J. (1992) Change in affective state as assessed by impaired colour-naming of anxiety-related words. *Current Psychology Research and Reviews*, in press.

Griffiths, R.R., Bigelow, G.E., and Liebson, I.A. (1986) Human coffee drinking: Reinforcing and physical dependence producing effects of caffeine. *Journal of Pharmacology and Experimental Therapeutics*, **239**, 416–25.

Griffiths, R.R., Bigelow, G.E., and Liebson, I.A. (1989) Reinforcing effects of caffeine in coffee and capsules. *Journal of Experimental Analysis of Behaviour*, **52**, 127–40.

Griffiths, R.R., Evans, S.M., Heishman, S.J., Preston, K.L., Sannerud, C.A., Wolf, B., and Woodson, P.P. (1990) Low-dose caffeine discrimination in humans. *Journal of Pharmacology and Experimental Therapeutics*, **252**, 970–8.

Griffiths, R.R. and Woodson, P.P. (1988) Reinforcing effects of caffeine in humans. *Journal of Pharmacology and Experimental Therapeutics*, **246**, 21–9.

Herman, C.P. and Polivy, J. (1975) Anxiety, restraint and eating behaviour. *Journal of Abnormal Psychology*, **84**, 666–72.

Hill, A.J. and Blundell, J.E. (1988) Role of amino acids in appetite control in man, in G. Huether (ed.) *Amino Acid Availability and Brain Function in Health and Disease*, Springer-Verlag, Berlin, pp. 239–48.

James, J.E. (1991) *Caffeine and Health*, Academic Press, London.

Jensen, A.R. (1987) Individual differences in the Hick paradigm, in P.A. Vernon (ed.) *Speed of Information-processing and Intelligence*, Ablex, Norwood New Jersey, pp. 101–75.

Johnson, L.C., Spinweber, C.L., Gomez, S.A., and Matteson, L.T. (1990a) Daytime sleepiness, performance, mood, nocturnal sleep: The effect of benzodiazepine and caffeine on their relationship. *Sleep*, **13**, 121–35.

Johnson, L.C., Spinweber, C.L., Gomez, S.A., and Matteson, L.T. (1990b) Benzodiazepines and caffeine: Effect on daytime sleepiness, performance and mood. *Psychopharmacology*, **101**, 160–7.

Kanarek and Swinney (1990) Effects of food snacks on cognitive performance in male college students. *Appetite*, **14**, 15–27.

Kraemer, H.C. (1981) Coping strategies in psychiatric clinical research. *Journal of Consulting and Clinical Psychology*, **49**(3), 309–19.

Larsen, R.J. and Ketelaar, T. (1989) Extraversion, neuroticism and susceptibility to positive and negative mood induction procedures. *Personality and Individual Differences*, **10**, 1221–8.

Leathwood, P.D. (1987) Tryptophan availability and serotonin synthesis. *Proceedings of the Nutrition Society*, **46**, 143–56.

Leathwood, P.D. and Pollet, P. (1983) Diet-induced mood changes in normal populations *Journal of Psychiatric Research*, **17**, 147–54.

Lieberman, H.R., Spring, B.J., Growdon, J.H., and Wurtman R.J. (1983) Mood, performance and pain sensitivity: Changes induced by food constituents. *J. Psychiatry Res.*, **17**, 135–45.

Lieberman, H.R., Spring, B.J., and Garfield, G.S. (1986) The behavioural effects of food components: strategies used in studies of amino acids, protein, carbohydrate and caffeine. *Nutrition Reviews*, **44** (Supplement), 61–70.

Lieberman, H.R., Wurtman, R.J., Emde, G.C., Roberts, C., and Coviella, I.L.G. (1987) The effects of low doses of caffeine on human performance and mood. *Psychopharmacology*, **92**, 308–12.

Liebert, R.M. and Morris, L.W. (1967) Cognitive and emotional components of test anxiety: A distinction and some initial data. *Physchological Reports*, **20**, 975–8.

Lipton, M.A., Nemeroff, C.B., and Mailman, R.B. (1979) Hyperkinesis and food additives, in R.J. Wurtman and J.J. Wurtman (eds) *Nutrition and the Brain*, Vol. 4, Raven Press, New York, pp. 1–27.

Loke, W.H. (1988) Effects of caffeine on mood and memory. *Physiology and Behaviour*, **44**, 367–72.

Macleod, C.M. (1991) Half a century of research into the Stroop effect: an integrative review. *Psychological Bulletin*, **109**, 163–203.

MacLeod, C.M., Mathews A., and Tata P. (1986) Attentional bias in emotional disorders. *Journal of Abnormal Psychology*, **95**, 15–20.

McNair D.M., Lorr M., and Doppelman L.F. (1971) *Profile of Mood States Manual*, Educational and Industrial Testing Service, San Diego.

Mandler, G. and Sarason, S.B. (1952) A study of anxiety and learning. *Journal of Abnormal and Social Psychology*, **47**, 166–73.

Mandler, G. and Watson, D.L. (1966) Anxiety and the interruption of behavior, in C.D. Spielberger (ed.) *Anxiety and Behavior*, Academic Press, New York.

Martens, R. (1969) Effect of an audience on learning and the performance of a complex motor skill. *Journal of Personality and Social Psychology*, **12**, 252–60.

Meisel, J.H. and Schlimme, E. (1990) Milk proteins: precursors of bioactive peptides. *Trends in Food Science and Technology*, **1**, 41–3.

Meunier, C. and Rule, B.G. (1967) Anxiety, confidence and conformity. *Journal of Personality*, **35**, 498–504.

Milon, H., Guidoux, R., and Antonioli, J.A. (1988) Physiological effects of coffee and coffee components, in R.J. Clarke and R. Macrae (eds) *Coffee, Volume 3: Physiology*, Elsevier Applied Science, London, pp. 81–124.

Morris, L.W., Davis, M.A., and Hutchings, C.H. (1981) Cognitive and emotional components of anxiety: Literature review and a revised Worry-Emotionality Scale. *Journal of Educational Psychology*, **73**, 541–55.

Ney, T. and Gale, A. (1988) A critique of laboratory studies of emotion with particular reference to psychophysiological aspects, in H.L., Wagner (ed.) *Social psychophysiology and emotion: Theory and clinical applications*, Wiley, Chichester, pp. 65–83.

Oatley, K. and Johnson-Laird, P.N. (1987) Towards a cognitive theory of emotions. *Cognition and Emotion*, **1**, 29–50.

Poliry, J. and Doyle, C. (1980) Laboratory induction of mood states through the reading of self-referent mood statements. Affective change or demand characteristics? *J. Abnormal Psychol.*, **89**, 286–90.

Power, R.P. and MacRae, K.D. (1977) Characteristics of items in the Eysenck Personality Inventory which affect responses when students simulate. *Brit. J. Psychol.*, **68**, 491–8.

Rogers, P.J. (1990) Why a palatability construct is needed. *Appetite*, **14**, 167–70.

Rogers, P.J. and Blundell, J.E. (1989) Separating the actions of sweetness and calories: effects of saccharin and carbohydrates on hunger and food intake in human subjects. *Physiology and Behaviour*, **45**, 1093–9.

Rogers, P.J., Keedwell, P., and Blundell, J.E. (1991) Further analysis of the short-term inhibition of food intake in humans by the dipeptide L-aspartyl-L-phenylalanine methyl ester (aspartame). *Physiology and Behaviour*, **49**, 739–43.

Ryan-Harshman, M., Leiter, L.A., and Anderson, G.H. (1987) Phenylalanine and aspartame fail to alter feeding behaviour, mood and arousal in men. *Physiology and Behaviour*, **39**, 247–53.

Sarason, I.G. and Stoops, R. (1978) Test anxiety and the passage of time. *Journal of Consulting and Clinical Psychology*, **46**, 102–9.

Saravis, S., Schachar, R., Zlotkin, S., Leiter, L.A., and Anderson, G.H. (1990) Aspartame: effects on learning, behaviour, and mood. *Paediatrics*, **86**, 75–83.

Schachter, S. (1964) The interaction of cognitive and physiological determinants of emotional state, in L. Berkowitz (ed.) *Advances in Experimental Social Psychology*, Vol. 1, Academic Press, New York.

Scheier, M.F. and Carver, C.S. (1977) Self-focussed attention and the experience of emotion: Attraction, repulsion, elation and depression. *J. Personality Soc. Psychol.*, **35**, 625–36.

Shanahan, M.P. and Hughes, R.N. (1986) Potentiation of performance-induced anxiety by caffeine in coffee. *Psychological Reports*, **59**, 83–6.

Sirota, A.D. and Schwartz, G.E. (1982) Facial muscle patterning and lateralization during elation and depression imagery. *J. Abnormal Psychol.*, **89**, 286–90.

Smith, A. (1988) Effects of meals on memory and attention, in M.M. Gruneberg, P.E. Morris, and R.N. Sykes (eds) *Practical Aspects of Memory: Current Research and Issues*, Vol. 2, pp. 177–482.

Smith, A.P. (1989) Diurnal variations in performance, in A.M. Colley, and J.R. Beech (eds) *Acquisition and Performance of Cognitive Skills*, Wiley, Chichester, pp. 301–25.

Smith, A.P. and Miles, C. (1986a) Acute effects of meals, noise and nightwork. *British Journal of Psychology*, **77**, 377–87.

Smith, A.P. and Miles, C. (1986b) Effects of lunch on selective and sustained attention. *Neuropsychobiology*, **16**, 117–20.

Smith, A.P., Leekam, S., Ralph, A., and McNeill, G. (1988) The influence of meal composition on post-lunch changes in performance efficiency and mood. *Appetite*, **10**, 195–203.

Smith, A.P., Rusted, J.M., Eaton-Williams, P., Savory, M., and Leathwood, P. (1990) Effects of caffeine given before and after lunch on sustained attention. *Neuropsychobiology*, **23**, 160–3.

Smith, A.P., Ralph, A., and McNeill, G. (1991a) Influences of meal size on post-lunch changes in performance efficiency, mood and cardiovascular function. *Appetite*, **16**, 85–91.

Smith, A.P., Rusted, J.M., Eaton-Williams, P., Savory, M., and Hall, S.R. (1991b) The effects of caffeine, impulsivity and time of day on performance, mood and cardiovascular function. *Journal of Psychopharmacology*, **5**, 120–8.

Spring, B (1986) Effects of foods and nutrients on the behaviour of normal individuals, in R.J. Wurtman and J.J. Wurtman (eds) *Nutrition and the Brain*, Vol. 7, Raven Press, New York, pp. 1–47.

Spring, B. (1990) Dietary selection, snacks and overeating in individuals characterised by carbohydrate preference and dysphoric mood. *Current Therapeutics Supplement*, **31**, 22–25.

Spring, B.J., Maller, O., Wurtman, J., Digman, L., and Cozolino, L. (1983) Effects of protein and carbohydrate meals on mood and performance: interactions with sex and age. *Journal of Psychiatric Research*, **17**, 155–67.

Steele, C.M. and Josephs, R.A. (1990) Alcohol myopia. *American Psychologist*, **45**, 921–33.

Stern, K.N., Chait, L.D., and Johanson, C.E. (1989) Reinforcing and subjective effects of caffeine in normal human volunteers. *Psychopharmacology*, **98**, 81–8.

Stroop, J.R. (1935) Studies of interference in verbal reactions. *J. Exp. Psychol.*, **18**, 643–62.

Teff, K.L., Young, S.N., and Blundell, J.E. (1989) The effect of protein or carbohydrate breakfasts on subsequent plasma amino acid levels, satiety and nutrient selection in normal males. *Pharmacology, Biochemistry and Behaviour*, **34**, 829–37.

Teasdale, J.D. and Taylor, R. (1981) Induced mood and accessibility of memories: An effect of mood state or of mood induction procedure. *Brit. J. Clinical Psychol.*, **20**, 39–48.

Thayer, R.E. (1989) *The Biopsychology of Mood and Arousal*, Oxford University Press, New York.

Triesman, A.M. and Geffen, P. (1967) Selective attention: Perception or response? *Quarterly Journal of Experimental Psychology*, **19**, 1–18.

Velten, E. (1968) A laboratory task for the induction of mood states. *Behaviour Research and Therapy*, **6**, 473–82.

Watts, F.N., McKenna, F.P., Sharrock, R., and Trezise, L. (1986) Colour naming of phobia-related words. *British Journal of Psychology*, **77**, 97–108.

Warburton, D.M. and Rusted, J.M. (1991) Cholinergic systems and information processing capacity, in J. Weinman and J. Hunter (eds) *Memory: Neurochemical and Abnormal Perspectives*, Harwood, London, pp. 87–104.

Warburton, D.M. and Rusted, J.M. (1989) Memory assessment, in I. Hindmarch and P.D. Stonier (eds) *Human Psychopharmacology: Measures and Methods*, Vol. 2, Wiley, Chichester, pp. 155–178.

Weiskrantz, L. (1968) *Analysis of Behavioural Change*, Harper and Row, London.

Wurtman, R.J. and Wurtman, J.J. (1989) Carbohydrates and depression. *Scientific American*, **260**, 50–7.

Yerkes, R.M. and Dodson, J.D. (1908) The relation of strength of stimulus to rapidity of habit formation. *Journal of Comparative and Neurological Psychology*, **18**, 459–82.

Young, S.N. (1986) The effect on aggression and mood of altering tryptophan levels. *Nutrition Reviews*, **44** (Supplement), 112–22.

10 Consumer expectations and their role in food acceptance
A.V. CARDELLO

10.1 Introduction

10.1.1 Food acceptance: definition and measurement

The study of the human response to food is a complex and rapidly evolving field. It encompasses a wide range of scientific disciplines, ranging from food science and technology to nutrition, biochemistry, physiology, psychology, marketing and catering. As may be expected in such an interdisciplinary area, numerous scientific concepts have evolved to describe various aspects of the phenomenon under investigation. However, the terminology used to describe these concepts, as well as the methods for measuring them, differ from one discipline to another. Food 'acceptance' is one such concept. Since the focus of this chapter concerns factors that influence food acceptance, I would first like to describe and define food acceptance and then to detail the operational approach that we have used to measure it in the laboratory.

Figure 10.1 is a schematic model of human food-related behavior. At its most basic level, 'food' can be considered as a sensory stimulus, the physicochemical characteristics of which are determined by a variety of ingredient, processing and storage variables. The study of these variables and their effects on food falls within the domain of food science and technology. When an individual encounters food, its physicochemical characteristics interact with the human senses to produce experiences of its appearance, taste, smell, texture, etc. The theoretical and empirical study of the transformation of physicochemical energy into these basic human sensations defines the area of psychology known as 'psychophysics'. In applied areas of food science it is termed 'sensory evaluation'. At the next level of information processing, these basic sensory attributes are integrated with other biobehavioral and cognitive information. The sources of this higher-order information may include bodily states (hunger, thirst), learning and memory, psycho-social and cultural influences, and a variety of cognitive variables. Although each of these factors has been demonstrated to have significant effects on perception, psychophysicists and sensory scientists have not routinely addressed their

FOOD - RELATED BEHAVIORS

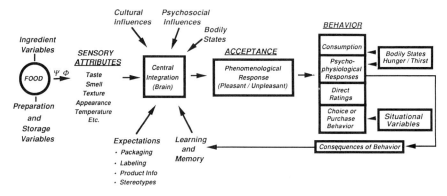

Figure 10.1 Schematic model of food-related behaviors.

effects. Rather, they have sought to minimize them through tight experimental controls, random sampling, and austere test conditions. The systematic study of these factors and their effect on food behavior has often been left to other disciplines, including those of psychobiology, social and cognitive psychology, nutrition, anthropology and consumer marketing.

The final product of the integration of basic sensory information with cognitive and other higher-order variables is a perception of the stimulus within a complex, contextual background. This is frequently accompanied by a concomitant emotional or hedonic response that falls along a continuum of 'pleasantness/unpleasantness' or 'like/dislike'. This hedonic response is what many investigators choose to call 'acceptance'. By its nature, it is a phenomenological experience. Experimental psychologists would call it an 'intervening variable', because its existence cannot be confirmed directly. In order to quantify or otherwise measure it, the observation of some behavioral response is required, as depicted in Figure 10.1. Regardless of the particular behavioral response that is measured, the sensory and hedonic experience of the food interacts with post-ingestional effects to produce consequences that feedback on learning, memory and bodily states. These, in turn, affect subsequent responses to that food item. This process is reflected by the feedback arrows in Figure 10.1.

The problem that the researcher faces when studying food acceptance is that, while the concept of acceptance is inherently rooted in phenomenology, it must be measured through behavior. Yet the particular behavior that one uses to measure it will greatly affect the interpretation of the phenomenological event itself. For example, physiologists might argue that the frequency of neuronal firing in the lateral hypothalamic medial forebrain bundle is the best index of the degree of pleasure elicited by the stimulus. Cognitive psychologists might argue that direct, introspective ratings of the degree of pleasantness/unpleasantness by the individual are the best measures, while behavioral psychologists would argue that the only entities worth measuring are overt behaviors, i.e. such things as choice, consumption, complaint behaviors, etc.

Yet each of these measures has shortcomings. Electrophysiological measures must invoke the assumption of 'psychophysical parallelism', i.e. that there is a one-to-one association between neuronal discharges and specific phenomenological experiences. Needless to say, most psycho-physiological measures are also highly impractical, except in the most remote and artificial of laboratory settings. Consumption, as a behavioral index, is heavily dependent upon hunger, thirst and other metabolic factors. The alternatives of choice and/or purchase behavior are highly contextually dependent. Moreover, in the real world, these dependent measures are greatly affected by price, availability and other socio-economic variables that are difficult to control. Direct ratings of food acceptance, while seemingly the least fraught with extraneous influences, are also a form of behavior, both verbal and numerical. As such, the manner in which individuals interpret words and use numbers will influence direct introspective ratings.

As a psychologist, the author considers the phenomenological aspects of food acceptance to be the most interesting and the most challenging of the many problems facing scientists in the field. There is a certain prima facie validity to an individual's self-report that he 'likes' or 'dislikes' a particular food item. As a psychophysicist, the author also believes that the phenomenology of food acceptance can be measured using subjects' direct self-reports, in the same way that subjects' direct self-reports of the perceived taste intensity of a model solution can be used as a valid dependent variable to relate to physicochemical characteristics of the stimulus solution. Thus, the predominant behavioral measure used in the author's research has been direct ratings of food acceptance. The most common operational measure has been a self-report of like–dislike, using a nine-point hedonic scale (Peryam and Pilgrim, 1957). The latter also has considerable practical appeal, because of the 35-year history of food acceptance data collected via this method in the author's laboratory and a predecessor laboratory at the Quartermaster Food and Container Institute in Chicago.

10.1.2 Overview

As mentioned previously, most psychophysicists and other sensory scientists spend considerable time controlling outside influences on their data. Many of these undesirable influences are environmental, inherent to the testing facility's physical layout, vicinity to other activities, lighting and air exchangers, etc. Other, more insidious influences are brought to the testing situation by the subject. These include physiological, cognitive, social and cultural influences. These influences are much more difficult to control. Not uncommonly, after many years of testing, the sensory scientist comes to find that the very factors that he sought to control become of greater interest than the sensory effects that he/she had previously sought to protect from these influences. So it is that the author's research has slowly shifted from controlling these 'non-sensory' influences to studying them.

The intent in this chapter is to focus attention on a cognitive variable that the author believes has significant impact on food acceptance and general sensory perception. It will begin with a review of certain pieces of data taken from published research that have contributed to the thought process leading to the thesis of this chapter. The review will focus on specific studies and data that suggest that a construct that may be referred to as 'expectations' can contribute significantly to our understanding of food acceptance. This construct will then be further developed with detail of some recent experiments in which expectations have been directly manipulated and their effects on sensory judgment and food acceptance observed.

10.2 The plausible role of expectations in food behavior

It should come as no surprise that the sensory attributes of a food play a significant role in its overall acceptance. Were that not the case, there would be wide-scale unemployment among sensory scientists currently working in the food industry. However, of greater interest to psychologists is uncovering general rules and principles that govern the role of these sensory attributes in food acceptance. For example, it has been well established that there are specific patterns to the growth of pleasantness/unpleasantness as a function of the intensity of food-related sensory attributes. Similarly, there is now a large body of data showing that certain tastes and odors are differentially preferred/rejected at birth. Models have also been constructed to account for changes in these innate preferences/aversions over time. These shifts in preference have been shown to occur through a variety of mechanisms, including mere exposure to previously novel or aversive foods, classical taste/odor aversion conditioning, and sensory specific satiety. In this section are reviewed some of these general principles as they relate to the prediction of food acceptance. However, the focus will be on selected data that suggest that consumers' 'expectations' about the sensory or hedonic properties of food can have as powerful an effect on perceived food acceptance as the actual physicochemical properties of the food itself.

10.2.1 Oral texture and temperature

On the whole, the effects of oral tactile sensations on acceptability have been less often studied than the effects of other sensory attributes, e.g. taste, odor, appearance. However, the role of texture on food acceptance has received increasing attention since the germinal work of Szczesniak and co-workers (1963, 1971, 1972), who examined consumers' awareness and attitudes toward various food textures. Although much of this research has confirmed that consumers suffer from a general lack of awareness and paucity of language for describing textural sensations, the importance of texture to food acceptance is aptly reflected in the large number of consumers who avoid such texturally unappealing products as squid, raw oysters, brains, liver and tapioca pudding. In fact, several studies have shown that texture is much more frequently cited as a reason for disliking a food than as a reason for liking it (Szczesniak, 1972; Sawyer *et al.*, 1988). Whether such reports are due to innate or acquired dislikes for the textural attributes of these products or may, in some cases, be due to preconceived expectations about the likely texture of these products, is something we will return to later.

Although studies have identified specific textural attributes as important sensory factors in the acceptability of a variety of foods (Hendrix *et al.*, 1963; Schutz *et al.*, 1972; Szczesniak and Kahn, 1971; Yoshikawa, *et al.*, 1970a,b,c; Okabe, 1979; Cardello *et al.*, 1983; Cardello and Maller, 1987; Szczesniak, 1991), the issue of consumer 'awareness' of food texture is critical to understanding its overall contribution to food acceptance. For example, in almost all studies that have been conducted with consumers, flavor is more frequently cited than texture as the reason for liking or disliking a food (see Jerome (1975) for a cultural exception with Afro-Americans). However, one common exception is bland foods, where texture, by default, is more likely to be the focus of consumer attention. The role of awareness or attention to texture can be seen in the data in Figure 10.2. These data are from a study of the relationships between perceptions of texture by naïve consumers and by texture profile panelists who have been trained to attend to the textural attributes of food (Cardello *et al.*, 1982). The data show acceptability ratings of bread products as a function of the instrumentally-determined texture of the bread. As can be seen, the trained panel ratings are much more greatly affected by the rheological variation in the products than are those of the naïve subjects. These results are supported by several other studies (Moskowitz *et al.*, 1974; Sawyer *et al.*, 1984, 1988) in which it has been

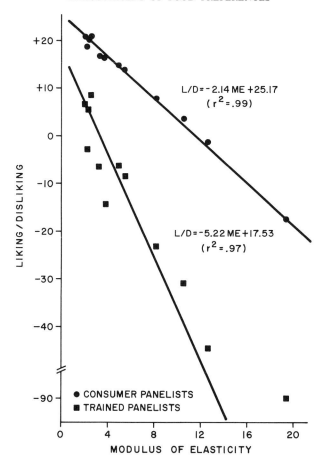

Figure 10.2 Liking/disliking ratings of bread products as a function of the modulus of elasticity (instrumental measure of texture) for both consumer and trained panelists (from Cardello *et al.*, 1982).

shown that the perceptual response range to textural variations in food is broadened with attribute-relevant training, and is consistent with earlier data of Szczesniak (1971) showing a greater awareness of texture among those who attend to food and food textures as part of their daily activities.

In a more recent study, consumer awareness of food geometry was examined by manipulating sensory and situational cues, e.g. sight of the food, manual contact with it, and sequential versus simultaneous presentation, that could aid the discrimination of differences in the size of food samples and associated textural judgments. The results of this study showed a linear relationship between the number of cues provided and the judged differences in hardness and chewiness of the samples (Cardello and Segars, 1989).

In one of the earliest papers on consumer awareness and/or attention to texture, the suggestion was first made that consumer 'expectations' may influence both the attention paid to texture and overall liking of the food. In this paper by Szczesniak and Kahn (1971) and quoted in Bourne (1982), the statement is made,

> If the texture of a food is the way people have learned to expect it to be, and if it is psychologically and physiologically acceptable, then it will scarcely be noticed. If, however, the texture is not as it is expected to be . . . it becomes a focal point for criticism and rejection of the food.

The critical part of this statement for the present discussion is the association of disconfirmed expectations with negative effect or acceptability. Although the statement is based on data from consumer interviews, the proposed relationship of disconfirmed expectation with decreased acceptance is consistent with other empirical observations.

A similar attribution of the effect of consumer expectations on texture perception has been made by Vickers (1991) to explain the occurrence of outliers in data relating perceived oral crispness and auditory crispness (Vickers, 1982). The outliers were responses to three foods in which oral crispness judgments did not correlate well with auditory crispness judgments. Judgments of oral crispness were lower than judgments based on their sound. The three foods were two types of humidified crackers and blanched celery. The explanation given was that

> judgments of crispness may have been affected by the subjects' expectation for the product. Blanched celery may have been much less crisp than the subjects expected it to be when they picked it up (Vickers, 1991, p. 92).

The implication here is that the judgment of crispness is dependent upon the level of expected crispness in the product and that products that are less crisp than expected are rated lower in oral crispness than products that confirm a given expectation of crispness.

The above quotations have been cited for two reasons. First is to document the fact that the use of the concept of 'expectations' to account for perturbations in both sensory and hedonic data is not new. Although these authors have used the term 'expectations' in a colloquial manner and only as a *post hoc* explanation of the observed data, as will be seen, investigators in other areas have also proposed this concept as an explanatory variable. The second reason is that the statements are relatively clear in focusing attention on the importance of *disconfirmation* of expectations. Disconfirmation and the direction of disconfirmation (better/worse) for hedonic expectations are important to the analysis of alternative cognitive mechanisms that may be responsible for these effects.

The other area of oral tactile sensation that is important for food acceptability is thermal perception. In general, there are three mechanisms by which temperature can affect food acceptance. First is the direct effect that temperature has on the molecular activity of the stimulus. This can produce increases or decreases in the rate that stimulus molecules interact with the receptor surface, thereby altering its sensory profile and, possibly, affecting acceptability. Second are the potential effects that the temperature of the food may have on receptor sensitivities themselves. Lastly are effects that operate through conditioned preferences for certain foods consumed at certain temperatures. Concerning the first two mechanisms, a body of data has accumulated on the effects of temperature on threshold and suprathreshold responses to sapid compounds (Stone et al., 1969; Pangborn, et al., 1970; Moskowitz, 1973; McBurney et al., 1973; Larson-Powers and Pangborn, 1978; Bartoshuk et al., 1982; Calvino, 1986). In a review of this area, Green and Frankmann (1987) concluded that, with the exception of a decrease in the perceived intensity of sucrose at lower temperatures, 'the effect of temperature on taste intensity is not a reliable phenomenon'. These investigators proposed that the lack of reliability was due to the failure to control the temperature of the tongue, not simply the temperature of the solutions, in these studies. In their own studies, Green and Frankmann showed that the temperature of the tongue exerts greater control over the perceived taste intensity of the solutions than does the temperature of the solution. From these data they concluded that temperature has a greater effect on the sensory transduction process than it has on the thermo-molecular properties of the solutions.

What exactly is the relationship between temperature and acceptability for various foods and beverages? Figure 10.3 shows data for the acceptability of thirteen foods and beverages as a function of temperature. As can be seen, most foods that are commonly served hot, e.g. entrée items, increase in acceptability from 40° to 140°F. On the other hand, foods or beverages that are normally served cold, e.g. milk and lemonade, decrease in acceptance with increasing temperature. Products that are served either hot or cold, e.g. coffee, show high acceptance at both temperature extremes, but low acceptance at room temperature (Cardello and Maller, 1982). Lester and Kramer (1991) have also shown that foods that are typically served hot are rated higher in acceptability and are consumed more when heated, as compared to when they are served at ambient temperature.

The differences in preferred temperatures for foods led Zellner et al. (1988) to a series of experiments that have also implicated consumer expectations as a factor in food acceptance. In these experiments it was shown that the acceptability of beverages served at different temperatures can be significantly altered by simply changing the subject's 'expectation'

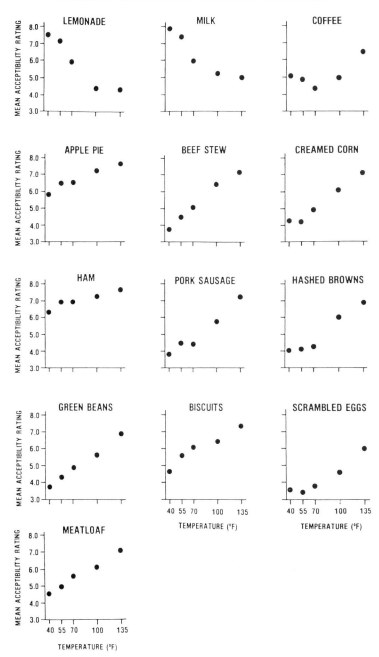

Figure 10.3 Acceptability ratings as a function of serving temperature for a variety of common foods and beverages (from Cardello and Maller, 1982).

concerning the temperature at which the beverage is typically consumed. In their study they used guanabana and tamarind juices that were served cold or at room temperature. One group of subjects was told that the juices are normally served at room temperature, while the other group was given no information. The data showed that the former group gave significantly higher acceptance ratings to the room temperature samples than did the latter group. In their conclusions, Zellner *et al.* state

> Tasting a beverage at an unfamiliar temperature can decrease the degree to which subjects report they dislike the beverage, at least temporarily. This indicates that the rejection of such beverages at certain temperatures is, at least in part, the result of expectations based on learned ideas of appropriateness. Expectations concerning the flavor of these substances at unfamiliar temperatures are worse than the actual experience of drinking them.
>
> Altering the expectations of the subject regarding which temperatures are appropriate and which are inappropriate can also change reports of liking for beverages at different temperatures. If subjects are led to believe that certain temperatures are appropriate for unfamiliar beverages they tend to report that they like them more at the temperatures they are told are appropriate.

Zellner *et al.*, in addition to focusing attention on the concept of 'expectations', also point to the relationship that may exist between the concepts of expectation and 'appropriateness'. The latter concept has been discussed by Schutz (1988) and the reader is referred to this treatment of the concept and its suggested role in food acceptance.

10.2.2   Flavor

While the texture of food products can have a profound effect on perceived acceptability, an even greater influence is exerted by the flavor of food. The first question that we can ask is 'are some tastes/odors innately preferred or rejected?' Certainly there is ample evidence showing that infants reject both bitter and sour tastes and accept sweet tastes (Nisbett and Gurvitz, 1970; Desor *et al.*, 1973, 1975; Lipsitt, 1977; Crook, 1978; Steiner, 1979; Rosenstein and Oster, 1988). In addition, some odors have been shown to be differentially preferred/rejected (Rosenstein and Oster, 1988). These early predispositions appear to remain strong throughout life, so that adult food cravings tend to be characterized by sweet tastes and pleasant smells, whereas aversions are frequently characterized by bitter tastes and foul smells (Blank and Mattes, 1990). However, innate preferences can be altered by experience and/or conditioning. These experimental and conditioning effects can decrease or increase acceptance. In the former case, a vast literature has evolved showing the effects of conditioned taste aversions (Garcia *et al.*, 1966; Garcia and Koelling, 1966) by the pairing of hedonically neutral tastes with illness induced by chemical and radiological means (see Garb and Stunkard, 1974; Berstein and Webster, 1980; Logue *et al.*, 1981; Pelchat and Rozin, 1982; and Bartoshuk and Wolfe, 1990 for representative studies with humans). In the case of preference conditioning, positive effects on acceptance have been found by simple repeated exposure to a novel taste/odor/food (Torrance, 1958; Capretta and Rawls, 1974; Domjan, 1976; Birch *et al.*, 1987; Davis and Porter, 1991), through flavor–flavor associations (Holman, 1975; Fanselow and Birk, 1982; Zellner *et al.*, 1983; Breslin *et al.*, 1990), and through pairing of flavors with post-ingestional satiety signals (Booth, 1972, 1981; Booth *et al.*, 1982; Tordoff *et al.*, 1987; Birch *et al.*, 1990). The conclusion to be drawn from these effects of learning and conditioning is that humans and other organisms have predispositions to accept or reject certain tastes and odors, however these predispositions are malleable and can be overridden by a variety of experiential factors.

Whether flavor preferences/aversions are innate or learned, it is obvious that the taste and odor of food has profound effects on its acceptability and consumption. The evidence of this fact is so pervasive in daily life that no purpose is served in documenting this point. However, there are several critical facts about the relationship between taste, odor and acceptability that are worthy of consideration here. For example, while it is well established that taste and odor intensity grow as a power function of physical intensity (Stevens, 1957), acceptability does not follow this or any other simple monotonic relationship. In the case of most sensory attributes that are acceptable throughout a broad range of their intensity continuum, e.g. sweetness, acceptability (pleasantness) increases with increasing physical intensity up to a certain point, whereupon pleasantness declines with further increases in intensity. The optimal level of acceptability is often referred to as the 'breakpoint' or 'bliss' point for that continuum (Moskowitz *et al.*, 1974). In the case of sensory attributes that are unpleasant throughout most of their sensitivity continuum, pleasantness declines monotonically with increasing concentration.

Of special interest to our present considerations is the fact that these relationships between sensory attributes and acceptability can be entirely reversed depending upon the context in which the flavor attribute appears. Take for example the data in Figure 10.4. Numerous psychophysical studies of sweetness have utilized sugar-in-water solutions as test stimuli. Consistent with the relationships just discussed, these studies have shown sweetness intensity to increase with increasing concentration, while the curve for pleasantness/preference increases up to a certain concentration and then flattens or decreases. In this study (Maller *et al.*, 1982), increasing concentrations of sucrose were tested in either a water solution (Figure 10.4a) or in eggs (Figure 10.4b). By following the pleasantness and preference ratings as a function of sucrose concentration, one sees that in water, the pleasantness and preference curves behave as expected, increasing to the point where they reach an asymptote and/or decline.

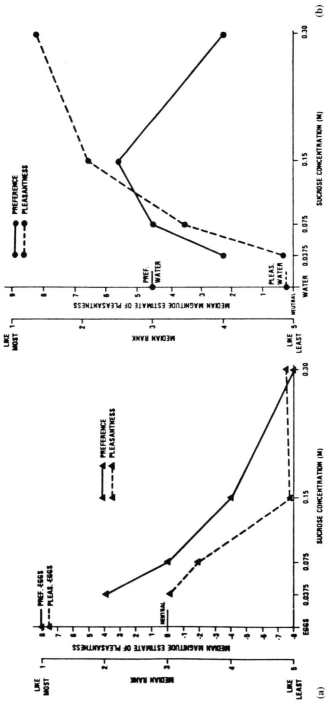

Figure 10.4 Preference and pleasantness as a function of sucrose concentration as judged in (a) scrambled eggs or (b) water solutions (from Maller *et al.*, 1982).

However, in eggs the same concentrations of sucrose result in pleasant-ness and preference curves that decline monotonically and then level off. Why should the pleasantness of sucrose behave so differently when per-ceived in eggs rather than water? One obvious answer is that sweetness is not normally associated with eggs. One might say that 'sweetness' is not an attribute that one 'expects' in eggs. The context in which tastes/odors are experienced is critical to the degree of pleasantness or unpleasantness that they elicit and no taste or odor can be said to elicit invariably pleasant or unpleasant sensations, without consideration of the context in which they are presented or the expectations that the context sets.

Studies of contextual effects in flavor perception have frequently focused on the effect of stimulus intensity ranges on the perception of the flavor intensity and pleasantness of test solutions. Most of these studies have manipulated the range of intensities or the frequency of presentation of two or more intensities in a series. Thus, Riskey et al. (1979) demon-strated that intensity ratings of the sweetness of fruit flavored drinks were increased when the samples were presented in a series of low sweet-ness samples, whereas the same samples were rated as less sweet when presented in a series containing higher intensity samples. Ratings of pleasantness of the samples were also affected by this contextual manipu-lation, with lower concentrations of sucrose judged to be more pleasant in the high intensity context and vice versa. McBride (1982, 1985) has shown similar effects for ratings of the sweetness and pleasantness of milk drinks and for the flavor intensity and pleasantness of fruit-flavored drinks presented in either high or low concentration series. Other effects re-ported by Riskey (1982) and Lawless (1983) for the saltiness of soups, Johnson and Vickers (1988) for the sweetness of lemonade, and Kroeze (1982) for the degree of suppression of saltiness/sweetness in NaCl/sucrose solutions, have all provided support to relativistic models of perception, e.g. adaptation-level theory (Helson, 1964) and range-frequency theory (Parducci, 1965; Poulton, 1968).

More recent research on contextual effects on flavor have begun to examine the role that the quality of the stimuli play in such effects. For example, Lawless (1989) and Lawless et al. (1991) have shown context dependent changes in the perception of odor quality. In these studies ambiguous odors of citrus/woody character were perceived as more woody when presented within the context of pure citrus odors, but more citrus-like when presented among pure woody odors. Marks, in several recent studies, has examined the role that the qualitative similarity between subsets of stimuli that differ in both quality and intensity has on con-textual intensity effects (Marks et al., 1986; Marks, 1988; Marks and Warner, 1991; Rankin and Marks, 1991). Using both auditory and flavor stimuli, these investigators have shown that the magnitude of contextual effects on intensity is a function of the qualitative similarity between the contextual stimuli and the test stimuli. For example, saltiness was decreased to a greater extent when presented within the context of a series of high intensity NaCl stimuli and low intensity sucrose solutions than when presented within the context of high intensity NaCl stimuli and low intensity NaCl/sucrose mixtures.

In the studies by Marks, the two stimulus subsets always differed in intensity ranges. However, in a somewhat obscure experiment conducted by Carlsmith and Aronson in 1963, a series of iso-intense solutions of sucrose and quinine sulfate was presented to subjects. These investigators were also looking for differential intensity effects as a function of stimulus quality. However, rather than manipulating intensity, these investigators manipulated the 'expected quality' of the stimulus. This was accomplished by providing the subjects with cues to the quality of the stimulus to be presented. In some cases the stimulus that was presented was consistent with the cue, i.e. sucrose was expected/sucrose was presented, in other cases it was inconsistent with the cue, i.e. sucrose expected/quinine presented. In all cases, judgments were made of the perceived intensity of the solutions. The results of this study showed significant differences in the perceived intensity of the solutions on trials when the subjects were given the stimulus that was not cued to them, as contrasted to trials when they were given the stimulus that was cued. However, consistent with the evolving theme of this chapter, this effect only occurred when the subject had a strong 'expectation' for the solution (defined by a criterion number of 'correct' trials preceding the target trial). Moreover, the effect on perceived intensity was different depending upon the quality of the stimulus, i.e. sucrose solutions which disconfirmed an expectancy were rated less sweet than sucrose solutions that confirmed an expectancy, but quinine solutions which disconfirmed an expectancy were rated more bitter than quinine solutions that confirmed an expectancy. At first glance, the differential intensity effects by quality appear inexplicable. However, the authors reconciled these data by proposing that, in both cases, the disconfirmed sensory expectations produced negative affect, a situation that would be reflected in both lower sweetness and higher bitterness ratings.

Carlsmith and Aronson's (1963) data were the first to draw a link between sensory perception of flavor, expectations, and affect (acceptability). The authors interpreted their results in terms of cognitive dissonance theory (Festinger, 1957), stating that 'if a person expects a particular event (X) and instead, a different event (Y) occurs, he will experience dissonance. Consequently, he will judge Y to be less pleasant than if he had had no previous expectancy'. While this interpretation is, in fact, consistent with dissonance theory, as we shall see, this is only one possible model to account for the results observed when product expectations are disconfirmed.

10.2.3   Appearance: food and its packaging

If one considers the numerous and varied ways in which consumers come into contact with food, it would be safe to conclude that the appearance of the food and/or its package constitutes the first sensory impression of the product. Appearance includes such basic sensory attributes of the food as its color, shape and size, as well as more complex attributes, such as translucency, gloss, or surface texture. Of all these visual aspects of food, the effect of color is the most dramatic, universal, and well-studied.

Some of the earliest experimental work on the effects of color on food perception and acceptance was conducted over a half-century ago by Moir (1936) and Dunker (1939), who first showed the strong association between a food and its color. Since that time numerous studies have shown the dramatic effects of color on taste recognition and taste intensity (Pangborn, 1960; Maga, 1974; Kostyla and Clydesdale, 1978; Johnson *et al.*, 1983; Christensen, 1983; Roth *et al.*, 1988), on flavor detection and identification (Dubose *et al.*, 1980; Urbanzi, 1982; Kanig, 1955; Hall, 1958) and on acceptability (Schutz, 1954; Worthington, 1960; Maga, 1973; SIK, 1976; Tuorila-Ollikainen *et al.*, 1984; Dubose *et al.*, 1980).

In our own laboratory, we have demonstrated important effects of color on the consumer perception and acceptability of a wide range of beverages, bakery products, meat and fish (Dubose *et al.*, 1980, 1981; Cardello *et al.*, 1983; Sawyer *et al.*, 1988). Perhaps most interesting for its possible relationship to consumer expectations are some data on inappropriate food colors. The data shown in Table 10.1 are taken from a study in which the stimuli consisted of three flavors of fruit drink (cherry, orange and lime) and a flavorless control (Dubose *et al.*, 1980). Each was prepared in a red, orange, green or colorless version using typical fruit-beverage color additives. Samples were presented in random order to subjects who were asked to identify the flavor of the beverage from a list of alternatives. As can be seen, the perception of the flavor identity of the beverage was significantly affected by the color of the beverage. As the color of the beverage changed from the one normally associated with its flavor to one not normally associated with it, a significant percentage of the perceived flavor identifications shifted from the 'correct' flavor to the flavor one would expect for that color. This effect is most evident in the flavorless sample, where the greatest percentage of flavor identity responses were for the flavor commonly associated with the beverage's color. In a similar but more recent study, inappropriately colored fruit-flavored beverages were found to both produce lower accuracy in flavor (odor) identification and result in reduced acceptance as compared to appropriately colored beverages (Zellner *et al.*, 1991).

In another often cited experiment (Wheatley, 1973), the effect of disconfirmed color expectations on food acceptance was more dramatically

Table 10.1 Data from a study of the effects of atypical colors on flavor identification of beverages. The cell entries are the percentages of consumer flavor responses, collapsed according to flavors with similar color associations, made to each beverage color/flavor combination. As can be seen, the color of the beverage increases the likelihood that the flavor will be perceived as being one that is normally associated with that color

Flavor response	Test beverage															
	Cherry flavored				Orange flavored				Lime flavored				Flavorless			
	Red	Orange	Green	Colorless	Red	Orange	Green	Colorless	Red	Orange	Green	Colorless	Red	Orange	Green	Colorless
Cherry/strawberry/raspberry	92.6	51.8	44.4	55.5	44.7	3.7	–	3.7	33.3	–	–	–	33.3	–	–	11.1
Orange/apricot	–	29.6	3.7	3.7	33.3	81.5	29.6	29.6	3.7	59.2	–	3.7	–	29.6	–	3.7
Lime/lemon–lime	–	3.7	37.0	14.8	11.1	3.7	40.7	22.2	33.3	22.2	85.1	66.7	–	–	40.7	–
Lemon/grapefruit/apple	3.7	7.4	3.7	18.5	–	3.7	14.8	29.6	3.7	11.1	3.7	14.8	14.8	14.8	11.1	22.2
Blueberry/grape/other	3.7	7.4	11.1	7.4	7.4	3.7	11.1	3.7	7.4	3.7	7.4	7.4	11.1	11.1	7.4	14.8
No flavor	–	–	–	–	7.4	3.7	3.7	7.4	18.5	3.7	3.7	7.4	40.7	44.4	40.7	48.1

Source: Dubois *et al.*, 1980.

demonstrated by having subjects eat a 'normal' meal under light-masking conditions. At a specified time during the meal, normal lighting was resumed, revealing to the subjects blue steak, green french fries, and red peas. Subjects were reported to have become nauseated from the sight of the food. Similar but less dramatic effects of inappropriate food colors frequently occur when naïve consumers encounter such common super-market products as white mint ice cream, brown eggs, brown (over-aged) meat, or green apples/bananas. Common explanations of these color effects on food acceptance range from innate neophobia (the item is rejected merely because it is novel) to learned associations between the inappropriate color and other negative qualities normally associated with that color in food, e.g. the case of unripened or spoiled foods. However, in keeping with the thesis presented in this chapter, it is suggested that many of these color effects can be explained as resulting from discon-firmed consumer expectations. That is, the common factor in most of these cases is that sensory expectations about the normal appearance of these foods have not been met. The normal flavor and texture of the meat, peas, and french fries that were consumed in the dark led to normal expectations about their color that were not met when the lights were turned on; the white color of the mint ice-cream led to expectations about its likely flavor (vanilla) that were not met; and the name, shape and other situational cues normally associated with the supermarket eggs, meat, and fruit all led to certain sensory expectations about the color of these products that were not met. In the case of the colored beverage studies, one can interpret the data of Dubose et al. (1980) as showing that the color of the beverages led to specific sensory expectations about their flavor. To the extent that the flavor was ambiguous, these expectations affected subjects' flavor responses in the direction of the expected flavor. Moreover, in the study by Zellner et al. (1991) the lower acceptability of inappropriately colored beverages is consistent with an interpretation of decreased affect under conditions of disconfirmed expectations. The mechanisms by which such effects on sensory perception and acceptance may occur is, as yet, unclear, but several alternative mechanisms will be discussed later in this chapter.

Although the visual appearance of the food itself is a powerful influence on its acceptability, so too is the visual appearance of its package, i.e. its shape, color, design, and associated logos, symbols, brand and item names (Hutchings, 1977). Since brand and item names are primarily ideational, they will be discussed separately. However, concerning the effects of package appearance, US military ration packages serve as an excellent model by which to examine the role of package-induced expec-tations, because ration packages have traditionally been of drab color with no distinguishing designs, logos, or brand names. In a recent set of studies aimed at developing more consumer-friendly packages for rations,

Figure 10.5 Three experimental ration packages designed with commercial-like graphics and the current military package design (center, top).

the effect of package appearance on food acceptance was studied. Four test packages were developed (Figure 10.5). Three of the packages (two zipper-sealed pouches of different colors/designs and a paperboard box) were designed with brighter, more attractive colors, and commercial-like designs. The fourth was a design copy of the existing military packaging for the MRE (Meal Ready-to-Eat) ration. All four packages were labeled with the acronym 'OPRS' to represent the name of a fictional ration system.

One hundred and eighty-three soldiers were shown the four packages in an incomplete block design and were asked to rate the packages on a variety of appearance and functionality attributes (Kalick and Cardello, 1991). Relevant to the issue of food expectations, soldiers were also asked to rate 14 attributes of the food contained inside the packages (without seeing or tasting them). Analysis of the data showed significant differences in the ratings assigned to the food products inside the packages. For instance, the food contained in both the zipper-sealed pouches and the paperboard box was perceived as better tasting, having higher quality ingredients, being more appetizing and being more likely to be made by a reputable company than the food in the standard MRE package. Soldiers also agreed that the commercial-like zipper-sealed packages and box were more likely to contain 'food I like' than the standard MRE. In addition, the food contained in the zipper-sealed pouches was perceived as significantly fresher tasting, easier to clean-up and more natural looking than food contained in either the box or the standard MRE. Since the test subjects did not actually taste or consume the food inside the package, what is the likely mechanism by which ratings of the food were affected? One explanation is that the brighter, more commercial-like packaging led subjects to expect better quality food and that their ratings reflected this higher expectation. A generalized 'halo effect' is also a possibility, where the novel packaging might be expected to elicit more positive ratings of any and all aspects of the ration and consumption situation.

In a second study, commercial packages and brand names were examined for their effect on ratings of both acceptability and food intake, not simply attitudes (Kramer *et al.*, 1989). In this study subjects consumed and rated the acceptability of a pudding served either in a plain white package, in one of two different military packages, or in its normal commercial package. Consumer ratings of the acceptability of the pudding and the total number of grams consumed were significantly higher when the pudding was packaged in its commercial brand package than in any of the other three packages.

In a third study, we took a slightly different approach and asked whether the differences in the acceptability of military versus commercial food would be affected by whether they were presented in military or commercial packages (Cardello *et al.*, 1985). Four food items were chosen, such that each was available in both a military-pack version that met military specifications and a commercial version that was a high quality, national brand leader. Both the military and commercial items were presented to different groups of subjects in either the military or commercial packages (packages were emptied and their contents interchanged and repackaged). Subjects were presented items in their test packages, they opened and tasted them, and rated them for acceptability. The data are shown in Figure 10.6. While no significant differences were found between the acceptability ratings of the military and commercial items when each was presented in military packages, a significant difference was found between the samples when served in the commercial packages. These latter results are somewhat perplexing. The reason is that, while the increased ratings of the acceptability of food served in more visually appealing packages can be easily interpreted as being the result of a generalized learning or 'halo' effect, i.e. positive associations with the package transfer to its contents, the results of this study do not lend themselves to such an interpretation. The reason is that such a 'halo' effect should have produced higher acceptability ratings for all foods presented in commercial packages. The intrinsic quality of the food should have made no difference. There is no justifiable basis by which to account for the differential effects seen in Figure 10.6. In order to account for such effects an explanation that takes into account the interaction of the food with the package is required. A model of packaging effects that is based on the expectations that the package elicits about the sensory and hedonic quality of the food and the degree to which those expectations are confirmed or disconfirmed by the food product can adequately

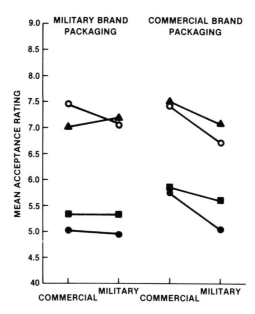

Figure 10.6 Mean acceptability ratings for four food items produced by commercial versus military vendors when presented in either commercial or military packaging. ▲, Grape jelly; ○, salted crackers; ■, non-dairy creamer; ●, instant coffee (from Cardello *et al.*, 1985).

account for these results. That is, high expectations for product quality induced by commercial packaging are met only by high quality foods. Lower expectations induced by military packaging are adequately met by both military food and by commercial foods. If consumer acceptability is a function of the degree of confirmation/disconfirmation of expectations, then the fact that both foods confirmed the minimal quality expectations set by the military packages, but only the commercial food met the quality expectation set by the commercial packages, would adequately account for the differential effects seen in the two conditions.

The implications of an 'expectation' model for food packaging and product marketing are far-reaching. From a strategic marketing standpoint it means that one must make a careful assessment of the product's ability to deliver on the key product elements that are touted to the public through packaging and other forms of advertising and communication. If such communications set realistic expectations that are met by the product, consumer satisfaction will not be detrimentally affected. However, high expectations of quality that are not met by actual product characteristics may lead to varying degrees of discontent. As shall be shown, any adequate model of the effects of disconfirmed expectations should be able to make accurate predictions for situations in which

1. high expectations are disconfirmed by product attributes;
2. low expectations are disconfirmed by product attributes;
3. both high and low expectations are confirmed.

10.2.4 Ideational effects

The foregoing sections have identified several areas of research within the sensory evaluation literature that are amenable to the interpretation that preconcieved expectations about the sensory or hedonic properties of food can affect subsequent perceptions of these properties. Such expectations can be generated by a variety of ideational or cognitive elements associated with the food. For example, when one goes into a restaurant and looks at the menu, one sees a list of food names, usually followed by a short description of the item that is designed to communicate the basic sensory properties of the item, e.g. 'fried in a light batter', 'cooked to a golden brown', 'tender pieces of juicy steak', etc. These descriptions, in combination with past experiences with the item name, create certain expectations about the likely sensory properties of the product, and, in turn, how much it will be liked. It is upon these cognitive data that one selects an item. Thus, if you order 'lasagna', you have certain expectations about what that lasagna will be like, in terms of the product attributes that are personally relevant to you, e.g. the type of sauce, the firmness/tenderness of the pasta, whether it will have meat or not, salt/ spice level, etc. These expectations are created by the item name, menu description, previous experience with lasagna in this restaurant, in other restaurants, at home, etc. The hypothesis being presented here is that your satisfaction, liking or acceptability for the lasagna that you receive, is a function, not only of the intrinsic sensory characteristics of the lasagna, but also of the degree to which the lasagna matches or mismatches your sensory and hedonic expectations. A lasagna that is lauded as 'gourmet' by a panel of esteemed chefs, but that does not meet your personal expectations, will not be liked as well as a less acclaimed lasagna that does meet your personal expectations.

The effects of ideational or cognitive stimuli on food acceptance are quite common, occurring with restaurant and institutional foods, as well as with branded, supermarket foods. In the case of institutional foods, expectations can be well ingrained and affect whole classes of foods. The data in Figure 10.7 are ratings of the *expected* acceptability of ten different food items when served in various foodservice settings (full service restaurant, fast food restaurant, airline, foodservice, etc.). The differences in ratings are extreme, from 3.5 for certain types of hospital food to greater than 8.0 for restaurant food. Moreover, the differences in expected acceptability between foodservice settings is constant across food items. Clearly, one must ask how these large differences in expected acceptabi-

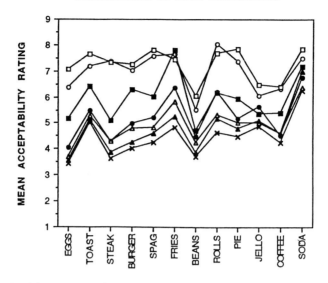

Figure 10.7 Mean 'expected' acceptability ratings for ten food and beverage items under seven possible foodservice operations. Data are in response to food names only. ○, Restaurant; ●, school; □, home; ■, diner/fast-food restaurant; △, military; ▲, airline; ×, hospital.

lities affect actual acceptability ratings. What if the same items were presented to consumers for actual tasting, but under situational conditions that led the consumer to believe he/she was eating airline versus commercial restaurant food? Moreover, what would be the effect on acceptability of serving food that was far better or worse than the established expectation?

In the case of supermarket foods, the studies mentioned previously on the effects of package design could be considered as resulting from ideational cues, especially when item and brand names are included in the packaging. Studies such as those by Pronko and Bowles (1949), Allison and Uhl (1964) Gacula *et al.* (1986) and Sheen and Drayton (1988) have demonstrated significant effects of brand-identity on the perception and rated acceptability of beer, soap products, hot dogs, and cola beverages. In the latter study it was demonstrated that simply labeling a high preference cola with its brand name will increase its acceptance over that given to the same cola in a blind taste test.

Ideational cues are also important in most sensory and food preference testing situations, where significant effort is expended on establishing and controlling sensory-related aspects of the experimental situation. However, many ideational cues are often left to chance or habit. Such cues as the name given to the item to be tested, its serving vessel, and product or user information can all have important effects on rated acceptability. In sensory studies conducted in our own laboratory, simply changing the name of a product from 'squid' to 'seafood tidbits' and from 'tofu' to 'oriental tidbits' resulted in significant effects on acceptability ratings. Providing certain types of product information, e.g. information on versatily of use, also had positive effects on stated purchase and use among subjects who had not previously tried the product (Cardello *et al.*, 1985). In a recent study we have examined the effect of panelist knowledge concerning the intended use and/or the intended user of a product being tested for acceptability in the laboratory. In the case of intended use, the subjects (200 military personnel of varied ages, 200 civilians under 40 years old and 200 civilians over 40 years old) rated five foods under three informational conditions. Subjects were told that this food would be eaten either (1) in a traditional consumption environment (restaurant, dining hall); (2) in a military field environment; or (3) in a military field environment but only by 18- to 25-year-old soldiers. For all subjects, acceptability ratings of the food were higher under the condition in which subjects thought the food was targeted for field use (Cardello *et al.*, 1991). Moreover, while there were no differences in acceptability ratings between conditions (2) and (3) for the under-40 population, the acceptability ratings were significantly lower in condition (3) for the '40 and over' age group. Clearly, in both circumstances, cognitive factors influenced acceptability ratings. Whether these subjects have lowered expectations for food served in the field than for food served in conventional settings, or if older individuals feel that younger persons have higher expectations for food, is not clear from the testing that was done. Nevertheless, it is clear that such ideational variables can have a significant impact on consumer ratings of food acceptability.

10.3   Consumer expectations and food acceptance

10.3.1   *Expectations as a construct*

The previous sections have raised the possibility that various results reported in the sensory and food acceptance literature are open to interpretation in terms of the subjects' 'expectations' and their subsequent confirmation or disconfirmation affecting both sensory and hedonic responses. However, most of these studies have merely used the term as a vague, *post hoc* explanatory variable. Expectations as a psychological construct has received no formal discussion and/or treatment in the sensory literature. In contrast to this situation, the concept of 'expectations' has received much greater attention in the fields of learning and social psychology, where it has played a central role in several cognitive theories of behavior. For example, Tolman (1938, 1951) used the term and the concept of 'expected consequences of behavior' to account for both animal and human learning. Tolman proposed that learned associations were the result of repeated pairings of stimuli that, eventually, cause one stimulus to lead to a belief or expectancy that the next stimulus will occur. In other learning situations, the stimulus leads to expectancy that a reward will result if a particular behavior is emitted in response to the stimulus. In the case of an animal being conditioned in a maze, the start box of the maze serves as a stimulus that leads to the expectancy that another stimulus (food) will be found at the end of the maze. In fact, Tolman's view of the initial stimulus (lights, sounds, etc. in the start box) was that it is perceived within a contextual background that includes the organism's past history of experiences with the stimulus. This view is very similar to contemporary stimulus-context views in perceptual psychology. The parallel between Tolman's analysis of animals running a maze for food and our considerations of the effect of expectations on human hedonic responses to food, is best seen when one considers what happens when the maze is correctly run and either no food is found in the goal box or a negative stimulus is found in the goal box. Under these conditions, the expectancy is disconfirmed. In learning theory these situations are referred to as 'extinction', whereupon the belief/expectancy on subsequent occasions is reduced or modified, and 'punishment', a negative hedonic experience that also produces a reduction in response stength. The parallels between Tolman's animals running mazes without 'expected' rewards and humans encountering products that fail to meet 'expected' standards are obvious.

Tolman was not the only psychologist to give attention to the concept of expectations. Meehl and MacCorquodale (1951), MacCorquodale and Meehl (1953), and MacCorquodale et al. (1954) gave the concept explanatory status in their theory of motivation, as did Rotter (1955) in his theory of social learning, and Atkinson (1954, 1957, 1958) in his theory of achievement motivation. These theories are now generally known as Expectancy × Value theories (see Feather, 1982, for a review). Along with Adaptation Level Theory (Helson, 1964) and Cognitive Dissonance Theory (Festinger, 1957), this class of theories places emphasis on the fact that the actual stimulus in any situation is the relationship between the objective stimulus and prior experience or context. Perception and overt behavior are both held to be determined by relativistic mechanisms.

10.3.2 Expectations: sensory versus hedonic

Although learning and social psychologists have utilized the construct of 'expectations', a closer examination of the construct and its potential use in sensory research is needed. In most sensory applications 'expectations' can be thought of as being of two general types: (1) a sensory-based expectation, i.e. a belief that the stimulus (food product, etc.) will possess certain sensory attributes, each at certain intensities, or (2) an hedonic-based expectation, i.e. a belief that the product will be liked/disliked to a certain degree. Examples of sensory expectations include those that are likely to operate in the restaurant 'menu' situation where specific product attributes are implied by the menu description, and in the studies of inappropriately colored foods and beverages, and expected food temperatures. Examples of hedonic expectations include those that are likely to occur in response to new package designs and brand labels, or when other 'ideational' stimuli elicit a general expectation for a good or poor product. In certain situations, both types of expectations may be elicited simultaneously. A mismatch between expected and actual sensory attributes or between expected and actual liking will result in 'disconfirmation'. In the case of disconfirmed hedonic expectations, the disconfirmation can be positive (the stimulus/product is better than expected) or negative (the stimulus/product is worse than expected).

10.3.3 Models of the effect of disconfirmed expectations

Working with the notion that there are two distinct types of expectations that consumers may have about food, the next question is how the confirmation or disconfirmation of these expectations affects food perception and acceptance. Although sensory scientists have given oblique reference to the concept of expectations, no attempts have been made to formalize predictive models of these effects. However, for many years, market researchers have addressed the question of how the failure to deliver on advertising promises affects consumer satisfaction with products. In fact, a number of predictive models have been proposed to explain the effect of disconfirmed expectations on consumer satisfaction/dissatisfaction with such varied consumer products as vacuum cleaners, ballpoint pens and restaurant services. While the interested reader is referred to Insko (1967), Oliver (1977a,b, 1980), Latour and Peat (1979) and Oliver and DeSarbo (1988) for reviews of this literature, the essential elements of these theoretical models and their application to sensory and food acceptance research follows.

Current theoretical treatments of the effect of disconfirmed expectations can be reduced to four distinct models. These are the assimilation model (Hovland et al., 1957; Sherif and Hovland, 1961; Olshavsky and Miller,

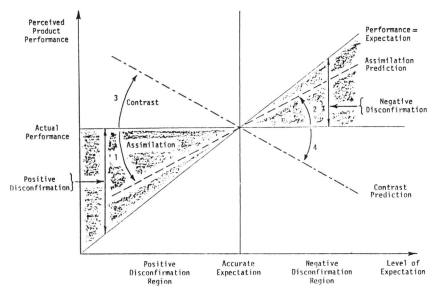

Figure 10.8 Assimilation and contrast model predictions of post-exposure product performance (from Oliver, 1977b).

1972; Olson and Dover, 1976, 1979), the contrast model (Hovland *et al.*, 1957; Sherif and Hovland, 1961; Dawes *et al.*, 1972), the assimilation-contrast model (Hovland *et al.*, 1957), and the generalized negativity model (Carlsmith and Aronson, 1963). These models can be differentiated on the basis of their predicted effects on perceived product performance in response to positive and negative disconfirmation. Figure 10.8 is taken from Oliver (1977b) and shows the specific predictions of the assimilation and contrast models. Predictions of the assimilation-contrast and generalized negativity models can also be inferred from this figure.

In Figure 10.8, the abscissa represents the consumer's level of expectation for the product and the ordinate represents perceived product performance, i.e. level of satisfaction/dissatisfaction or, in our case, level of liking/disliking. The diagonal line extending from the origin maps the points where product performance matches expectations. For some arbitrary product, 'actual performance' is represented by the horizontal line in the figure. For this product, expectation and product performance 'match' at the point of intersection with the diagonal line. Here, all models predict that perceived product performance will equal actual product performance. However, if actual performance is held constant and the level of expectation is varied, then positive disconfirmation (actual performance > expected performance) or negative disconfirmation (expected performance > actual performance) occurs, as indexed by the vertical distance between the level of actual performance and the point of intercept on the diagonal line.

Regardless of whether positive or negative disconfirmation occurs, the assimilation model predicts that perceived product performance will assimilate (become similar to) the level of expectation, as shown by arcs 1 and 2, respectively. Such a model might be used to account for the data from studies on food packaging and branding, where 'better' brands and 'better packaging' increase acceptability of the food, while generic or institutional brands and/or poor packaging decrease acceptability. The contrast model, on the other hand, predicts that perceived product performance will move in the direction opposite to the expectation. This effect is shown by arcs 3 and 4. A generalized model of this form does well in predicting the range-intensity context effects discussed earlier, where test stimuli undergo contrast effects and are judged as less intense within the context of a high intensity series, but more intense within the context of a low intensity series. Such a model can also be used to explain the hedonic effects reported by McBride (1985), in which acceptability was shown to be lowered when the sample was presented within a context of high acceptability samples and elevated when presented within a context of low acceptability samples.

The assimilation-contrast model is a hybrid form of the models just discussed. This model predicts that consumer satisfaction/dissatisfaction follows assimilation model predictions under conditions of low positive or low negative disconfirmation, but follows contrast model predictions under conditions of high positive or high negative disconfirmation. In other words, this model predicts that assimilation will occur when the actual sensory or hedonic attributes of the product differ only slightly to moderately from expectations. However, if the product differs significantly from expectations, the difference between expectation and reality becomes too large and a contrast effect occurs.

The last of the four models is the generalized negativity model. It predicts that perceived product performance decreases under all conditions of disconfirmation. This model is consistent with the cognitive dissonance model discussed earlier, and can account for such data as those of Carlsmith and Aronson (1963), in which negative affect occurs under all conditions of disconfirmed expectations.

10.3.4   Measuring expectations and confirmation/disconfirmation

In order to empirically study the effects of disconfirmed expectations on food acceptance, an operational definition of expectations and related constructs is needed. A casual examination of the kinds of verbal behavior that are commonly used in reference to food expectations can provide some insights. Statements that come to mind include: 'This ice cream is so creamy, you'll love it', 'My aunt's apple pie is the best you'll ever taste', 'If you like chewy cookies, taste these' and 'I hope the food here is as good as they say'. Such examples confirm the fact that some expectations are sensory based, e.g. *creamy* ice cream and *chewy* cookies, and some are purely hedonic, e.g. *'best* pie you'll taste' and 'food as *good* as they say'. Examples of post-consumption comments that reflect the confirmation/disconfirmation experience include 'It didn't have the right taste', 'It was done just right', 'It wasn't what I expected' or 'I didn't like it, it was too dry, salty, rich, . . . etc.'.

In our initial research on this concept and its role in food acceptance, we began by developing operational measures by which to index sensory and hedonic expectations and associated measures of confirmation/disconfirmation. The approach was to select measures that have operational validity based on well-founded roots in traditional sensory methodology. In the case of hedonic expectations, we have operationalized its definition as 'the expressed degree of anticipated liking for a future stimulus'. The method by which we have quantified it is through a direct rating of expected or anticipated liking using a traditional nine-point hedonic scale. For example, a subject may be posed the following situation and question: 'Shortly you will be presented with a serving of orange juice. How much do you expect to like it?' The response options given to the subject range from 1 = dislike extremely to 9 = like extremely, with 5 = neither like nor dislike. The methodology is parallel to standard hedonic scaling methods (Peryam and Pilgrim, 1957), except that the judgment is made prior to presentation of the stimulus and is a judgment of 'expected liking'. In the case of sensory expectations, the approach is somewhat different but consistent with established methods for assessing the sensory properties of food. In this case subjects are asked to generate an 'expected sensory profile' of the anticipated product. That is, subjects generate a series of intensity ratings for each of a variety of salient sensory attributes of the product. Subjects are asked to rate the 'expected or anticipated intensity of the following attributes of the product . . . '. A variety of intensity scales may be used, as long as provision is made for a 'zero' or 'not present' response.

Using these operationalized measures of sensory and hedonic expectations, measures of confirmation/disconfirmation were then developed using 'the degree of difference between expected and actual sensory or hedonic properties' as a definition of disconfirmation. In the case of hedonic disconfirmation, we have chosen the signed difference between expected and actual acceptability as a measure of positive and/or negative disconfirmation. For sensory disconfirmation we have used the unweighted average difference between the expected and actual attribute intensity ratings as a measure.

10.4 Experimental studies

10.4.1 Sensory disconfirmation

If sensory experience precedes hedonic response, then sensory disconfirmation must precede hedonic disconfirmation. However, once the hedonic response occurs, we can justifiably ask how sensory disconfirmation affects acceptability and how inferred measures of sensory disconfirmation and/or hedonic disconfirmation relate to direct post-stimulus ratings of confirmation/disconfirmation. Thus, in one of the first experiments in which we employed these measures, we examined the effect of sensory disconfirmation on acceptability and compared inferred measures of disconfirmation to a direct, self-reported measure of confirmation/disconfirmation (Cardello and Sawyer, 1992). The samples in this study consisted of a water soluble, edible film that had been used to coat a candy product. Thirty-eight consumers, who had no prior experience with edible films, served as the subjects. Subjects were informed about the uses of edible coatings in foods, but no information was provided about the sensory properties of the coatings. Subjects were instructed on how to produce an expected sensory profile of the edible coating and rated the expected intensity of nine salient sensory attributes that had been selected during pilot tests. In addition to judging the expected sensory attributes of the product, subjects rated the expected acceptability of the product.

After making these judgments, subjects were presented the coated product and were asked to generate a 'perceived' sensory profile, using the same set of nine attributes. Subjects also

1. judged the acceptability of the coating;
2. estimated the degree to which the edible coating matched/mismatched their initial expectations;
3. rated their likelihood of purchasing the product.

Subjects used a seven-point scale that varied from 1 = 'did not match my expectations' to 7 = 'matched my expectations perfectly' to directly rate the degree of confirmation/disconfirmation. The inferred measure of sensory disconfirmation was calculated as the unweighted mean of the absolute differences between the intensity ratings for the nine sensory attributes on the 'expected' and 'perceived' profiles.

Figure 10.9 is a plot of the difference between expected acceptability of the edible coating and its rated acceptability after tasting (post-test minus expected) as a function of the inferred measure of sensory disconfirmation. As sensory disconfirmation increased, post-test acceptability decreased relative to expected acceptability. Thus, when the sensory attributes of the product did not differ from expectations, judged acceptability was equal or higher than expected acceptability. However, when the product

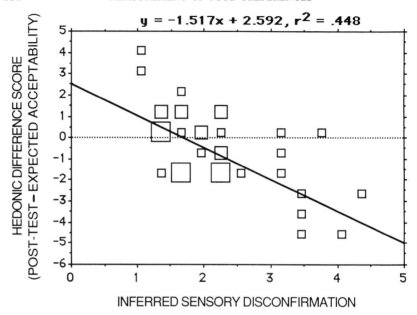

Figure 10.9 Linear regression plot of the hedonic difference score (post-test minus expected acceptability) as a function of inferred sensory disconfirmation (from Cardello and Sawyer, 1992).

attributes differed from expectations, rated acceptability was lower than expected acceptability.

If we stop to consider the measure plotted on the ordinate in Figure 10.9, one can see that it is somewhat analogous to our definition of hedonic disconfirmation. The difference is that the post-test rating of acceptability actually incorporates the presumed effect of any hedonic disconfirmation experience. In other words, the construct of hedonic disconfirmation refers to the difference between the expected acceptability of the product and the intrinsic acceptability of the product prior to any influence by expectation effects. Such would be the case if we had prior baseline acceptability ratings for products for which we subsequently manipulated expectations. However, in the present circumstance, where the products were completely novel to the subjects, no such measure was possible. The post-test measure of acceptability already reflects any effects of sensory or hedonic expectations. Nevertheless, the measure plotted on the ordinate of Figure 10.9 is a pertinent measure to compare with direct ratings of the overall disconfirmation experience, since it may well be this judgment of acceptability that is compared cognitively to expected acceptability in order to arrive at an overall judgment of the degree to which the product matched or mismatched expectations. In point of fact, the correlation between this measure and the direct judgment of the degree to which the product matched/mismatched expectations was found to be higher ($r = 0.65$; $p < 0.01$) than the correlation between sensory disconfirmation and the direct judgment of disconfirmation ($r = 0.47$; $p < 0.05$).

Both the inferred sensory measure of disconfirmation and subjects' direct judgments of disconfirmation were also found to be negatively correlated with product acceptance and with purchase intent. For both measures, greater disconfirmation resulted in lower acceptance and reduced purchase intent, supporting the general hypothesis that disconfirmed expectations result in negative affect and reductions in associated behavioral responses.

10.4.2   Direct manipulation of expectations

In the above experiment, expectations were not manipulated directly. Rather, the expectations were merely those that the individual brought to the testing situation. However, if the construct of expectations and our operational measures of it are to be of utility, they should respond to direct manipulations that commonly operate in real world situations. Moreover, by directly manipulating expectations, it should be possible to produce both positive and negative disconfirmation of sufficient range and magnitude to allow assessment of the effects in terms of the predictive models described in the previous section.

In a second set of experiments (Cardello and Sawyer, 1992) consumer expectations were directly manipulated through information presented about the product. The information was designed to manipulate both sensory and hedonic expectations and to produce both positive and negative hedonic disconfirmation. The test product was a commercial pomegranate juice that had been adjusted with distilled water and sucrose to yield a product having neutral hedonic tone, i.e. a consumer acceptance rating of approximately 5.0 and a bitterness intensity rating of approximately 3.0 ('slightly bitter' on a 7-point intensity scale). One hundred and eight consumers were divided randomly into four groups. In order to establish different levels of expectation (and disconfirmation), the groups were differentially exposed to positive, negative, accurate or minimal product information immediately prior to the test. They were instructed as follows:

All groups Today you will be testing a sample of juice from a new kind of tropical fruit . . .

Groups 1–3 The juice was nationally tested with a large group of consumers last December. Almost everyone who tasted it said they . . .

Group 1 (control group: accurate expectation – confirmed) 'neither liked nor disliked' it. It had an average score of 5.0 on a 9-point scale and had average bitterness.

Group 2 (low expectation – positive disconfirmation) 'disliked it very much'. It had an average score of 1.9 on a 9-point scale and was described as 'very bitter'.

Group 3 (high expectation – negative disconfirmation) 'liked it very much.' It had an average score of 8.1 on a 9-point scale and was described as 'not bitter at all'.

Group 4 (expectation/disconfirmation not manipulated). These subjects were told only that they would be tasting a new kind of juice. No other information was provided.

After exposure to the product information, subjects in each group generated an expected sensory profile of the juice by rating the intensity that they *expected* for the sweetness, bitterness, sourness, fruit flavor, and astringency of the juice. They also rated how much they *expected* to like/dislike the juice. A sample of juice was then served to each panelist, and they were asked to generate a perceived sensory profile of the product using the same sensory attributes as before. Judgments of acceptability were also obtained.

The relationships between mean ratings for expected and perceived acceptance, and for expected and perceived intensity of bitterness are shown in Figure 10.10(a) and (b). Mean ratings of expected acceptability were significantly different among the experimental groups (Figure 10.10a), as were the mean bitterness ratings (Figure 10.10b), with the exception of Groups 3 and 4.

Since the test sample was formulated to have a pre-test acceptability of ~5.0, presentation of this sample to the various groups resulted in different levels of operationally defined confirmation/disconfirmation. For example, Group 3 had a mean expected acceptability of ~7.0, therefore, presentation of the test juice to this subject group would result in negative disconfirmation. Similarly, since Group 2 had a mean expected acceptability of ~3.0, positive disconfirmation would result in this group. In the control group (expected acceptability ~5.0), no disconfirmation would occur, while in Group 4 (expected acceptability ~6.3), a slight negative disconfirmation would occur.

The results of this study showed that the mean perceived acceptance rating in Group 3 was significantly higher than that for Group 1 (control), supporting an assimilation model effect for the high expectation–negative disconfirmation group. However, the mean ratings of perceived acceptability for both Groups 2 and 4 were not significantly different from the control, leading to the conclusion that positive disconfirmation and even intermediate levels of negative disconfirmation had no observable effect on overall ratings of acceptability.

Examination of the sensory expectation data in Figure 10.10(b) reveals a still more interesting set of results. Although the sensory disconfirma-

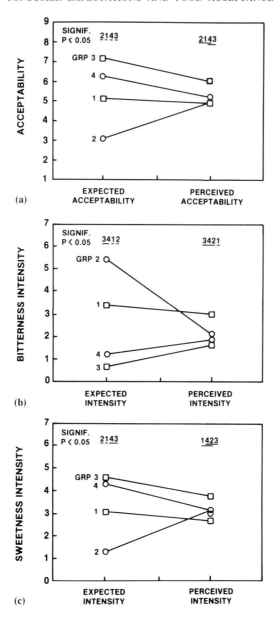

Figure 10.10 Mean ratings for expected and perceived acceptance (a), and for expected and perceived intensity of bitterness (b), and sweetness (c) in a modified pomegranate juice. Data show assimilation effects for acceptability and sweetness in Group 3, assimilation effects for bitterness in Groups 3 and 4, and a contrast effect for bitterness in Group 2 (from Cardello and Sawyer, 1992).

tion that was established in Groups 3 and 4 (the juice tasted more bitter than expected), resulted in an assimilation effect, i.e. the perceived intensity ratings were significantly lower than for Group 1, the high bitterness intensity expectation (Group 2) resulted in a pronounced contrast effect, i.e. a significant lowering of the perceived intensity rating below that of Group 1. Of particular interest from a sensory standpoint are the data in Figure 10.10(c). Although expectations were not manipulated for 'sweetness', 'fruit flavor', 'astringency', or 'sourness' (all of which were rated for both expected and perceived intensity), strong associative effects were observed nevertheless for each of these attributes. (Note the significant differences in Figure 10.10(c) among the sweetness ratings for the groups in spite of the fact that nothing was communicated to the groups about the likely sweetness of the juice.) In addition, the high sweetness intensity expectation that was induced in Group 3 resulted in a significant assimilation effect on perceived sweetness (Figure 10.10(c), Group 3 versus Group 1). This effect was attributed to the strong positive association of sweetness intensity with hedonic response and/or negative association of sweetness intensity with bitterness intensity. When subjects expect juice to be well-liked and low in bitterness they also assume it will be sweet; whereas when they expect it to be low in acceptance and bitter, they also assume it will lack sweetness. It seems that expectations derived from information about one or more product characteristics can influence expectations for other characteristics of the product as well. Moreover, the expectations established in this way result in similar disconfirmation effects. Regardless of how 'expectations' are formed, they can be confirmed or disconfirmed, and the resultant effects on product perception follow the same cognitive rules.

Of some additional interest is the fact that the ratings of expected acceptability, bitterness, and sweetness for Group 4 are intermediate to those of Groups 1 and 3. These results were interpreted to mean that, in the absence of contradictory information, consumers expect fruit juice to have relatively high acceptance and low bitterness. As such, it would be anticipated that the effects for Group 4 would be parallel and intermediate to those of Groups 1 and 3. The data in Figure 10.10(a)–(c) support this conclusion. These data also suggest that consumers have expectations about the hedonic and sensory properties of foods/beverages, independent of any active communication of information about the product to them. Moreover, these expectations appear to have similar effects on subsequent perception and acceptability as those that are established by direct product information.

Although the acceptability data shown in Figure 10.10(a) support an assimilation model of the effect of disconfirmed expectations under conditions of high expectation/negative disconfirmation, the bitterness data in Figure 10.10(b) show both assimilation (Groups 3 and 4) and contrast

(Group 2) effects. As pointed out previously, the assimilation–contrast model predicts that assimilation will occur under conditions of low disconfirmation, and that contrast will occur when disconfirmation is large. It may be that the level of disconfirmation produced in Groups 3 and 4 fell within the limits in which an assimilation effect would occur, but that the disconfirmation in Group 2 was sufficiently large to produce a contrast effect.

Unfortunately, examination of the ratings of expected bitterness for Groups 3 and 4 versus Group 2 do not support the contention that greater disconfirmation was produced by the stimulus for Group 2. However, an important point needs to be made here. The levels of confirmation/disconfirmation established in the above experiment were dependent upon the use of a single test product presented to all subjects. The actual acceptability and bitterness of this product was indexed in pilot studies with a separate, random group of consumers. Thus, individual variation in sensory sensitivities and/or preferences toward the product characteristics would likely introduce variability into the levels of sensory and/or hedonic disconfirmation experienced by individual subjects. Such a 'group approach' to the stimulus problem makes it difficult to accurately index and compare slight differences in the levels of disconfirmation experienced by different groups of subjects. In order to reduce this variability, the test product presented to each subject would have to be pre-tested with that subject prior to the start of the experiment in order to establish its acceptability and perceived sensory characteristics for purposes of indexing the actual degree of disconfirmation relative to expectations.

10.4.3   Direct manipulation of disconfirmation levels

In order to gain better control over disconfirmation levels affecting individual subjects, pre-test measures of acceptability were obtained in an experiment utilizing cola beverages as test stimuli (Cardello and Sawyer, 1992). The products consisted of six national and local brands of cola beverage. In a baseline screening test, 281 subjects judged the acceptability of each of the six cola beverages. From these data, 180 subjects were selected in accordance with the experimental design shown in Table 10.2. The six different groups were established so that expectations for the beverage to be served were either high, intermediate, or low. This was accomplished by instructing subjects that they were to receive and evaluate a cola that was found by 'a national survey of cola drinkers' to be indistinguishable from a specific brand of cola that they knew and with which they were familiar. Thus, each individual expected the cola beverage to be indistinguishable from a brand in the preliminary test that he/she had either disliked (rated 1, 2, or 3; Groups 1 and 6); liked (rated

Table 10.2 Experimental design used for cola beverage study (from Cardello and Sawyer, 1992)

| Group | Hedonic rating* in preliminary test | | Level of expectation | Level of disconfirmation |
	Brand expected	Brand tasted		
1	1, 2, 3	7, 8, 9,	Low	Large +
2	7, 8, 9	1, 2, 3	High	Large −
3	4, 5, 6	7, 8, 9	Intermediate	Intermediate +
4	4, 5, 6	1, 2, 3	Intermediate	Intermediate −
5	7, 8, 9	7, 8, 9	High	Low or none
6	1, 2, 3	1, 2, 3	Low	Low or none

* 9-point hedonic scale; 1 = dislike extremely, 9 = like extremely.
Source: Cardello and Sawyer, 1992.

7, 8, or 9; Groups 2 and 5); or been neutral to (rated 4, 5, or 6; Groups 3 and 4).

To ensure the desired levels of confirmation/disconfirmation, the beverage presented to each subject was one that that individual had either liked (rated 7, 8, or 9; Groups 1, 3, and 5) or disliked (rated 1, 2, or 3; Groups 2, 4, and 6) in the preliminary test. Subjects in Groups 5 and 6 (confirmation groups) were given the same brand of cola that they had been led to expect.

Before tasting the test beverage, subjects rated expected acceptability. They were then served an unlabeled sample of cola in accordance with the experimental design layed out in Table 10.2. Subjects rated how much they actually liked/disliked the beverage and gave a direct rating of perceived disconfirmation, i.e. whether the cola tasted better, worse, or the same as expected.

Table 10.3 shows (1) the mean baseline pre-test acceptability ratings for the brands of cola that subjects were served, (2) subjects' mean ratings of the 'expected' acceptability of the test cola after being told the 'brand' to expect, and (3) the mean acceptability rating of the cola when presented as the 'new' cola. With the exception of Group 6, the manipulations were effective in establishing the desired levels of expectation/disconfirmation. In Group 6, the low expectation/low disconfirmation manipulation did not have its intended effect, i.e. it did not produce an expected acceptability similar to the baseline acceptability. The higher level of expected acceptability in this group caused it to serve as another intermediate negative disconfirmation condition.

All groups showed assimilation of acceptability ratings toward expected levels. That is, the mean rating of acceptability for the beverages in the high expectation groups were significantly higher than in the preliminary

Table 10.3 Results of the cola beverage study. The assimilation effects in each group are reflected in the statistically significant shifts in acceptability (post-test versus pre-test) toward the 'expected' acceptability

Group	Pre-test acceptability	Expected acceptability	Level of disconfirmation	Post-test acceptability rating	t-Values post-test versus pre-test acceptance
1	7.8	3.8	Large +	6.4	$-4.33**$
2	2.4	6.1	Large −	6.0	$9.12**$
3	7.8	5.3	Intermediate +	6.1	$-6.24**$
4	2.2	5.3	Intermediate −	5.5	$9.83**$
5	7.8	6.5	Low/no disconfirmation	7.4	$-2.46*$
6	2.2	5.2	[a] Intermediate −	5.6	9.20

9-point hedonic scale; 1 = dislike extremely, 9 = like extremely.
[a] Note that this experimental condition was intended to produce low/no disconfirmation, using a low baseline acceptability beverage (see text).
$*p < 0.05$; $**p < 0.001$.
Source: Cardello and Sawyer, 1992.

test. Likewise, the mean ratings in the low expectation groups were significantly lower than in the preliminary test.

Figure 10.11 is a plot of the change in product rating for each subject as a function of the degree of disconfirmation he/she experienced, where disconfirmation was indexed by the algebraic difference in pre-test acceptability ratings for the expected and tasted brands. The strong positive association seen in Figure 10.11 reflects the fact that subjects who expected a worse product, rated the product lower than they had in the preliminary test, whereas subjects who expected a better product, rated it higher than in the preliminary test.

Under all conditions of positive and negative disconfirmation in this experiment, support was found for an assimilation model of disconfirmed expectations. Of course, the assimilation/contrast model predicts that contrast would occur only if the levels of disconfirmation are sufficiently high. However, cola beverages elicit a high degree of brand loyalty and preference. In the high positive and high negative disconfirmation conditions of this experiment subjects were told to expect their highest or lowest preference brands, but they were actually given the exact opposite. The resultant disconfirmation would be expected to be extremely high. However, no contrast effect was observed; only assimilation effects.

A Pearson product–moment correlation conducted between the inferred measure of disconfirmation i.e. rating of expected liking minus rating of perceived liking and direct ratings of disconfirmation, i.e. their response to the question 'Did the new cola taste as you had expected it to taste?' produced a high correlation coefficient ($r = 0.81$, $p < 0.001$). The

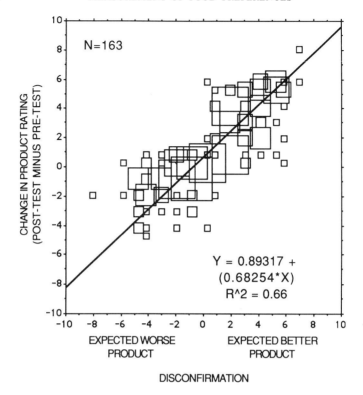

Figure 10.11 Linear regression plot of the change in acceptance rating (9-point hedonic scale) of cola beverages as a function of level of disconfirmation (from Cardello and Sawyer, 1992).

high correlation supports the notion that consumers' evaluation of the disconfirmation experience results from a comparison of perceived acceptability with expected acceptability.

With the exception of the one contrast effect that resulted from a high bitterness expectation, the results of the studies conducted to date lend support for an assimilation model of the effect of disconfirmed consumer expectations on product acceptance. The fact that the contrast effect occurred in only one of several high disconfirmation conditions across the several studies suggests that contrast effects may occur only rarely. This rarity of contrast effects is consistent with previous reports (Anderson, 1973; Olson and Dover, 1976). Further study is needed in order to identify the specific conditions that make contrast effects more likely, before any definitive model of disconfirmed expectations can be proposed.

10.5 Summary/conclusions

This chapter has put forth the hypothesis that disconfirmed consumer expectations play a significant role in both sensory perception of food and the determination of food acceptance. The evidence for this is drawn both from previously published data in the sensory and food acceptance literature that are amenable to such an interpretation, as well as from recent experiments that have been designed to examine specific aspects of the proposed effects. The chapter has also sought to review alternative models of the predicted effects of disconfirmed expectations and to suggest operational definitions and quantitative measures by which these theoretical constructs can be measured and the models tested using sensory research paradigms.

The empirical data collected to date have only touched the surface of this complex problem. Clearly, assimilation effects predominate, but contrast effects have also been observed. It is far too early to exclude any model from consideration. Much more needs to be done in order to understand the specific circumstances under which assimilation and contrast effects occur. Much more also needs to be done to understand the relationship between sensory disconfirmation and hedonic disconfirmation, and to determine if the effects observed on acceptance extend to other behavioral measures, e.g. choice, purchase and/or consumption.

The implications of the effects of disconfirmed consumer expectations are far-reaching. For strategic marketers and advertisers they raise the issue of the degree to which product expectations can be raised without risking severe failure if the product does not live up to those expectations. For those in institutional foodservice they raise serious questions about the best strategies for overcoming negative expectations about institutional food. Lastly, for the sensory scientist, these effects raise the issue of proper experimental methodology to control for the effects of subject's expectations about the food samples to be served. For all researchers involved in food acceptance and food behavior, the recognition, measurement and understanding of the role of consumer expectations should be of critical concern.

References

Allison, R.I. and Uhl, K. (1964) Influences of beer brand identification on taste perception. *J. Marketing Res.*, **1**, 36–9.

Anderson, R.E. (1973) Consumer dissatisfaction: The effect of disconfirmed expectancy on perceived product performance. *J. Marketing Res.*, **10**, 38–44.

Atkinson, J.W. (1954) Explorations using imaginative thought to assess the strength of human motives, in *Nebraska Symposium on Motivation*, Vol. 2, M.R. Jones (ed.), University of Nebraska Press, Lincoln, pp. 56–112.

Atkinson, J.W. (1957) Motivational determinants of risk-taking behavior. *Psychol. Rev.*, **64**, 359–72.

Atkinson, J.W. (1958) Toward experimental analysis of human motivation in terms of motives, expectancies, and incentives, in J.W. Atkinson edn., *Motives in Fantasy, Action, and Society*, Van Nostrand, Princeton, pp. 288–305.

Bartoshuk, L.M., Rennert, K., Rodin, J., and Stevens, J.C. (1982) Effects of temperature on the perceived sweetness of sucrose. *Physiol. and Behav.*, **28**, 905–10.

Bartoshuk, L.M. and Wolfe, J.M. (1990) Conditioned 'Taste' Aversions in Humans: Are They Olfactory Aversions? Paper presented at the meeting of the Association for Chemoreception Sciences, Sarasota, FL. April, 1990.

Bernstein, I.L. and Webster, M.M. (1980) Learned taste aversions in humans. *Physiol. and Behav.*, **25**, 363–6.

Birch, L.L., McPhee, L., Steinberg, L., and Sullivan, S. (1990) Conditioned flavor preferences in young children. *Physiol. and Behav.*, **47**, 501–5.

Birch, L.L., McPhee, L., Pirok and Steinberg, L. (1987) What kind of exposure reduces children's food neophobia? *Appetite*, **9**, 171–8.

Blank, D.M. and Mattes, R.D. (1990) Exploration of the sensory charateristics of craved and aversive foods. *J. Sens. Stud.*, **5**(3), 193–202.

Booth, D.A. (1972) Conditioned satiety in the rat. *J. Comp. Physiol. Psychol.*, **81**, 457–71.

Booth, D.A. (1981) The physiology of appetite. *Br. Med. Bull.*, **37**, 135–40.

Booth, D.A., Mather, P., and Fuller, J. (1982) Starch content of ordinary foods associatively conditions human appetite and satiation, indexed by intake and eating pleasantness of starch–paired flavors. *Appetite*, **3**, 163–84.

Bourne, M.C. (1982) *Food Texture and Viscosity: Concept and Measurement*, Academic Press, New York.

Breslin, P.A.S., Davidson, T.L., and Grill, H.J. (1990) Conditioned reversal of reactions to normally avoided tastes. *Physiol. and Behav.*, **47**, 535–8.

Calvino, A.M. (1986) Perception of sweetness: the effects of concentration and temperature. *Physiol. and Behav.*, **36**, 1021–8.

Capretta, P.J. and Rawls, L.H. (1974) Establishment of a flavor preference in rats: Importance of nursing and weaning experience. *J. Comp. and Physiol. Psychol.*, **86**, 670.

Cardello, A.V. and Maller, O. (1982) Acceptability of water, selected beverages and foods as a function of serving temperature. *J. Food Sci.*, **47**(5), 1549–52.

Cardello, A.V., Maller, O., Kapsalis, J.G., Segars, R.A., Sawyer, F.M., Murphy, C., and Moskowitz, H.R. (1982) Perception of texture by trained and consumer panelists. *J. Food Sci.*, **47**(4), 1186–97.

Cardello, A.V., Secrist, J., and Smith, J. (1983) Effects of soy particle size and color on the sensory properties of ground beef patties. *J. Food Qual.*, **6**, 139–51.

Cardello, A.V., Maller, O., Bloom-Masor, H., Dubose, C., and Edelman, B. (1985) Role of consumer expectancies in the acceptance of novel foods. *J. Food Sci.*, **50**, 1707–14, 1718.

Cardello, A.V. and Maller, O. (1987) Sensory texture analysis, an integrated approach to food engineering, in H.R. Moskowitz (ed.). *Food Texture: Instrumental and Sensory Measurement*, Marcel Dekker, Inc., New York, 177–215.

Cardello, A.V. and Segars, R.A. (1989) Effects of sample size and prior mastication on texture judgements. *J. Sens. Stud.*, **4**, 1–18.

Cardello, A.V., Sawyer, F.M., Kalick, J., and Lesher, L. (1991) Cognitive factors controlling food acceptance. Presented at 8th World Congress of Food Science and Technology, Toronto, Canada, September 29–October 4, 1991.

Cardello, A.V. and Sawyer, F.M. (1992) The effects of disconfirmed consumer expectations on food acceptance. *J. Sensory Stud.*, **7**, 253–77.

Carlsmith, J.M. and Aronson, E. (1963) Some hedonic consequences of the confirmation and disconfirmation of expectancies. *J. Abnorm. Soc. Psychol.*, **66**(2), 161.

Christensen, C.M. (1983) Effects of color on aroma, flavor and texture judgements on food. *J. Food Sci.*, **48**, 787–90.

Crook, C.K. (1978) Taste perception in the newborn infant. *Infant Behav. Dev.*, **1**, 52–69.

Davis, L.B. and Porter, R.H. (1991) Persistent effects of early odor exposure on human neonates. *Chem. Sens.*, **16**(2), 169–74.

Dawes, R.M., Singer, D., and Lemons, F. (1972) An experimental analysis of the contrast effect and its implications for intergroup communication and the indirect assessment of attitude. *J. of Personality and Social Psych.*, **21**(3), 281–95.

Desor, J.A., Maller, O., and Turner, R.E. (1973) Taste in acceptance of sugars by human infants. *J. Comp. Physiol. Psychol.*, **84**, 496–501.

Desor, J.A., Maller, O., and andrews, K. (1975) Ingestive responses of human newborns to salts sour and bitter stimuli, *J. Comp. Physiol. Psychol.*, **84**, 966–70.

Domjan, M. (1976) Determinants of the enhancement of flavored-water intake by prior exposure. *J. Exp. Psychol.*, **2**(1), 17.

Dubose, C.N., Cardello, A.V., and Maller, O. (1980) Effects of colorants and flavorants on identification, perceived flavor intensity, and hedonic quality of fruit-flavored beverages and cake. *J. Food Sci.*, **45**, 1393–9, 1415.

Dubose, C.N., Cardello, A.V., and Maller, O. (1981) Factors affecting the acceptability of low-nitrite smoked, cured ham. *J. Food Sci.*, **46**(2), 461–3.

Dunker, K. (1939) The influence of past experience upon perceptual properties. *Am. J. Psychol.*, **52**, 255.

Fanselow, M. and Birk, J. (1982) Flavor–flavor associations induce hedonic shifts in taste preference. *Animal Learn. and Behav.*, **10**, 233–8.

Feather, N.T. (ed.) (1982) *Expectations and Actions: Expectancy-Value Models in Psychology*, Lawrence Erlbaum Associates, Hillsdale, New Jersey.

Festinger, L. (1957) *A Theory Of Cognitive Dissonance*, Row & Peterson, Evanston, IL.

Gacula, Jr., M.C., Rutenbeck, S.K., Campbell, J.F., Giovanni, M.E., Gardze, C.A., and Washam II, R.W. (1986) Some sources of bias in consumer testing. *J. Sens. Stud.*, **1**, 175–82.

Garb, J.L. and Stunkard, A. (1974) Taste aversions in man. *Am. J. Psychiat.*, **131**, 1204–7.

Garcia, J., Ervin, F.R., and Koelling, R.A. (1966) Learning with prolonged delay of reinforcement. *Psychon. Sci.*, **5**, 121–2.

Garcia, J. and Koelling, R.A. (1966) Relation of cue to consequence in avoidance learning. *Psychon. Sci.*, **4**, 123–4.

Green, B.G. and Frankmann, S.P. (1987) The effect of cooling the tongue on the perceived intensity of taste. *Chem. Sens.*, **12**, 609–19.

Hall, R.L. (1958) Flavor study approaches at McCormick and Company, Inc., in *Flavor Research and Food Acceptance*, Reinhold, New York.

Helson, H. (1964) *Adaptation-level theory.* Harper and Row, New York.

Hendrix, J., Baldwin, R., Rhodes, V.J., Stringer, W.C., and Nauman, H.D. (1963) Consumer acceptance of pork chops. *Missouri Univ. Agr. Expt. Sta. Res. Bull.* no. 834.

Holman, E. (1975) Immediate and delayed reinforcers for flavor preferences in rats. *Learning and Motiv.*, **6**, 91–100.

Hovland, C.I., Harvey, O.J., and Sherif, M. (1957) Assimilation and contrast effects in reactions to communication and attitude change. *J. Abn. Soc. Psychol.*, **55**, 244–52.

Hutchings, J.B. (1977) The importance of visual appearance of food to the food processor and the consumer. *J. Food Qual.*, **1**, 267–78.

Insko, C.A. (1967) *Theories of Attitude Change*, Appleton-Century-Crofts, New York.

Jerome, N.W. (1975) Flavor preferences and food patterns of selected U.S. and Caribbean blacks. *Food Technol.*, **29**(6), 46.

Johnson, J.L., Dzendolet, E., and Clydesdale, F.M. (1983) Psychophysical relationships between sweetness and redness in strawberry flavored drinks. *J. Food Protect.*, **46**, 21–5.

Johnson, J.L. and Vickers, Z.M. (1988) Avoiding the centering bias or range effect when determining an optimum level of sweetness in lemonade. *J. Sens. Stud.*, **2**, 283–92.

Kalick, J. and Cardello, A.V. (1991) Consumer-oriented package designs: Improving military ration acceptance. Paper presented at Meeting of the Institute of Food Technologists, Dallas, TX, 1–5 June, 1991.

Kanig, J.L. (1955) Mental impact of colors in foods studied. *Food Field Reporter*, **23**, 57.

Kostyla, A.S. and Clydesdale, F.M. (1978) The psychophysical relationship between color and flavor of some fruit-flavored beverages – Influence of the addition of red. Unpublished manuscript.

Kramer, F.M., Edinberg, J., Luther, S., and Engell, D. (1989) The impact of food packaging on food consumption and palatability. Paper presented at Association for Advancement of Behavior Therapy, Washington DC, November, 1989.

Kroeze, J.H. (1982) The influence of relative frequencies of pure and mixed stimuli on mixture suppression in taste. *Percept. and Psychophys.*, **31**(3), 27(-8.

Larson-Powers, N. and Pangborn, R.M. (1978) Paired comparison and time intensity measurements of the sensory properties of beverages and gelatins containing sucrose or synthetic sweetners. *J. Food Sci.*, **43**, 41-6.

Latour, S.A. and Peat, S.A. (1979) Conceptual and methodological issues in consumer satisfaction research, in W.L. Wilkie (ed.) *Advances in Consumer Research*, Association for Consumer Research, **6**, 431-7.

Lawless, H. (1983) Contextual effect in category ratings. *J. Test. and Eval.*, **11**, 346-9.

Lawless, H.T. (1989) Exploration of fragrance categories and ambiguous odors using multidimensional scaling and cluster analysis. *Chem. Sens.*, **14**, 349-60.

Lawless, H.T., Glatter, S., and Hohn, C. (1991) Context-dependent changes in the perception of odor quality. *Chem. Sens.*, **16**, 349-60.

Lester, L.S. and Kramer, F.M. (1991) The effects of heating on food acceptability and consumption. *J. Foodservice Systems*, **6**, 69-87.

Lipsitt, L.P. (1977) Taste in human neonates: Its effect on working and heart rate in J.M., Weiffenbach, (ed.), *Taste and Development: The Genesis of Sweet Preference*, US Government Printing Office, Washington DC, pp. 124-42.

Logue, A.W., Ophir, I., and Strauss, K.E. (1981) The acquisition of taste aversions in humans. *Behav. Research and Therapy*, **19**, 319-33.

MacCorquodale, K. and Meehl, P.E. (1953) Preliminary suggestions as to a formalization of expectancy theory. *Psychol. Rev.*, **60**, 55-63.

MacCorquodale, K., Meehl, P.E., and Tolman, E.C. (1954) in W.K. Estes, S. Koch, K. MacCorquodale, P.E. Meehl, C. Muller, W. Schoenfeld and W.S. Verplanck, (eds), *Modern Learning Theory*. Appleton-Century-Crofts, New York, pp. 177-266.

Maga, J.A. (1973) Influence of freshness and color on potato chip sensory preference. *J. Food Sci.*, **38**, 1251.

Maga, J.A. (1974) Influence of color on taste thresholds. *Chem. Sens. and Flav.*, **1**, 115.

Maller, O., Cardello, A.V., Sweeney, J., and Shapiro, D. (1982) Psychophysical and cognitive correlates of discretionary usage of table salt and sugar by humans, in J.E. Steiner and J.R. Granchrow (eds.), *Proceedings of the Fifth European Chemoreception Research Organization Symposium*, IRL Press, London, pp 205-18.

Marks, L.E., Szczesiul, R., and Ohlett, P. (1986) On the cross-modal perception of intensity. *J. Exp. Psychol. Hum. Percept. Perform.*, **12**, 517-34.

Marks, L.E. (1988) Magnitude estimation and sensory matching. *Percept. Psychophys.*, **43**, 511-25.

Marks, L.E. and Warner, E. (1991) The slippery context effect and critical bands. *J. Exp. Psychol. Hum. Percept. Perform.*, 986-6.

McBride, R.L. (1982) Range bias in sensory evaluation. *J. Food Technol.*, **17**, 405-10.

McBride, R.L. (1985) Stimulus range influences intensity and hedonic ratings of flavor. *Appetite*, **6**, 125-31.

McBurney, D.H., Collins, V.B., and Glantz, L.H. (1973) Temperature dependence of human taste responses. *Physiol. and Behav.*, **11**, 89-94.

Meehl, P.E. and MacCorquodale, K. (1951) Some methodological comments concerning expectancy theory. *Psychol. Rev.*, **58**, 230-3.

Moir, H.C. (1936) Some observations on the appreciation of flavour in foodstuffs. *Chem. Ind.*, **55**, 145.

Moskowitz, H.R. (1973) Effects of solution temperature on taste intensity in humans. *Physiol. and Behav.*, **10**, 289-92.

Moskowitz, H.R., Kluter, R.A., Westerling, J., and Jacobs, H.L. (1974) Sugar sweetness and pleasantness, evidence for different psychological laws. *Science*, **184**, 583-5.

Nisbett, R. and Gurwitz, S. (1970) Weight, sex, and the eating behavior of human newborns. *J. Comp. Physiol. Psychol.*, **73**, 245-53.

Okabe, M. (1979) Texture measurement of cooked rice and its relationship to the eating quality. *J. Texture Stud.*, **10**, 131-52.

Oliver, R.L. (1977a) Effect of expectation and disconfirmation on postexposure product evaluations: an alternative interpretation. *J. Appl. Psychol.*, **62**(4), 480-6.

Oliver, R.L. (1977b) A theoretical reinterpretation of expectation and disconfirmation effects on postexposure product evaluations: experience in the field, in R.L. Day, (ed.), *Consumer Satisfaction, Dissatisfaction and Complaining Behavior* Papers from Marketing Research Symposium, School of Business, Indiana Univ., Bloomington, April 20–22, pp. 2–9.

Oliver, R.L. (1980) Theoretical bases of consumer satisfaction research: review, critique, and future direction, in C.W. Lamb, Jr. and P.M. Dunne, (eds), *Theoretical Developments in Marketing* American Marketing Association, Chicago, pp. 206–10.

Oliver, R.L. and DeSarbo, W.S. (1988) Response determinants in satisfaction judgments. *J. Consum. Res.*, **14**, 495–507.

Olshavsky, R.W. and Miller, J.A. (1972) Consumer expectation, product performance, and perceived product quality. *J. Mark. Res.*, **9**, 19–21.

Olson, J.C. and Dover, P. (1976) Effects of expectation creation and disconfirmation on belief elements of cognitive structure, in B.B. Anderson, (ed.) *Advances in Consumer Research* Association for Consumer Research 3, pp. 168–175.

Olson, J.C. and Dover, P. (1979) Disconfirmation of consumer expectations through product trial. *J. Appl. Psychol.*, **64**, 179–89.

Pangborn, R.M. (1960) Influence of color on the discrimination of sweetness. *Amer. J. Psychol.*, **73**(2), 229.

Pangborn, R.M., Chrisp, R.B., and Bertolero, L.L. (1970) Gustatory, salivary, and oral thermal responses to solutions of sodium chloride at four temperatures. *Percept. Psychophys.*, **10**, 289–92.

Parducci, A. (1965) Category judgement: A range frequency model. *Psychol. Rev.*, **72**, 407–18.

Pelchat, M.L. and Rozin, P. (1982) The special role of nausea in the acquisition of food dislikes by humans. *Appetite*, **3**, 341–51.

Peryam, D. and Pilgrim, F. (1957) Hedonic scale method of measuring food preferences. *Food Technol.*, **11**, 9.

Poulton, E.C. (1968) The new psychophys: six models for magnitude estimation. *Psychol. Bull.*, **69**, 1–19.

Pronko, N.H. and Bowles, J.W. (1949) Identification of cola beverages III: a final study. *J. Appl. Psychol.*, **33**, 605–8.

Rankin, K.M. and Marks, L.E. (1991) Differential context effects in taste perception. *Chem. Sens.*, **16**, 617–29.

Riskey, D.R., Parducci, A., and Beauchamp, G.K. (1979) Effects of context in judgements of sweetness and pleasantness. *Percept. and Psychophys.*, **26**, 171–6.

Riskey, D.R. (1982) Effects of context and interstimulus procedures in judgements of saltiness and pleasantness. *Selected Sensory Methods: Problems and Approaches to Measuring Hedonics*. ASTM STP 773, J.T. Kuznicki, Am. Soc. Testing and Materials, 71–83.

Rosenstein, D. and Oster, H. (1988) Differential facial responses to four tastes in newborns. *Child Dev.*, **59**, 1555–68.

Roth, H.A., Radle, L., Gifford, S.R., and Clydesdale, F.M. (1988) Psychophysical relationships between perceived sweetness and color in lemon and lime flavored beverages. *J. Food Sci.*, **53**, 1116–9.

Rotter, J.B. (1955) The role of the psychological situation in determining the direction of human behavior, in *Nebraska Symposium on Motivation* (M.R. Jones ed.), University of Nebraska Press, Lincoln, pp. 245–69.

Sawyer, F.M., Cardello, A.V., Prell, P.A., Johnson E.A., Segars, R.A., Maller, O., and Kapsalis, J. (1984) Sensory and instrumental evaluation of snapper and rockfish species. *J. Food Sci.*, **49**, 727.

Sawyer, F.M., Cardello, A.V., and Prell, P.A. (1988) Consumer evaluation of the sensory properties of fish. *J. Food Sci.*, **53**(1), 12–18, 24.

Schutz, H.G. (1954) Color in relation to food preference, in K.T. Farrell, J.R. Wagner, M.S. Peterson, and G.M. McKinnery (ed.), *Color in Foods: A Symposium*, p. 21.

Schutz, H.G., Damrell, J.D., and Locke, B.H. (1972) Predicting hedonic ratings of raw carrot texture in sensory analysis. *J. Texture Stud.*, **3**, 227–32.

Schutz, H.G. (1988) Beyond preference: appropriateness as a measure of contextual acceptance of food, in *Food Acceptability* (D.M.H. Thomson ed.), Elsevier Applied Science, London, 115–34.

Sheen, M.R. and Drayton, J.L. (1988) Influences of brand label on sensory perception, in *Food Acceptability* (D.M.H. Thomson ed.), Elsevier Applied Science, London, 89–99.

Sherif, M. and Hovland, C.I. (1961) *Social Judgement: Assimilation and Contrast Effects in Communication and Attitude Change*, Yale University Press, New Haven.

SIK Information (English) (1976) The taste of decolored foods. **5**(1), 6. From Marit Neymark (1975) Sa fargas var smak, In Rad och Ron, **10**, 16.

Steiner, J.E. (1979) Human facial expressions in response to taste and smell stimulation, in H.W. Reese and L.P. Lipsett, (eds), Vol 13 *Adv. Child Dev.* Academic Press, New York, 257–95.

Stevens, S.S. (1957) On the psychophysical law. *Psych. Rev.*, **64**, 154–81.

Stone, H., Oliver, S., and Kloehn, J. (1969) Temperature and pH effects on the relative sweetness of suprathresholds mixtures of dextrose and fructose. *Percept. Psychophys.*, **5**, 257–60.

Szczesniak, A.S. (1971) Consumer awareness of texture and of other food attributes. II. *J. Text. Stud.*, **2**, 196–206.

Szczesniak, A.S. (1972) Consumer awareness of and attitudes to food texture. II: Children and teenagers. *J. Texture Stud.*, **3**, 206–217.

Szczesniak, A.S. (1991) Textural perceptions and food quality. *J. Food Quality*, **14**(1), 75–86.

Szczesniak, A.S. and Kahn, E.L. (1963) Consumer awareness of texture and other food attributes. *Food Technol.*, **17**, 74–7.

Szczesniak, A.S. and Kahn, E.L. (1971) Consumer awareness of and attitudes to food texture. I: Adults. *J. Texture Stud.*, **2**, 280–95.

Tolman, E.C. (1938) The determiners of behaviors at a choice point. *Psychol. Rev.*, **45**, 1–41.

Tolman, E.C. (1951) *Behavior and Psychological Man*. University of California Press, Berkeley.

Tordoff, M.G., Tepper, B.J., and Friedman, M.I. (1987) Food flavor preferences produced by drinking glucose and oil in normal and diabetic rats: evidence for conditioning based on fuel oxidation. *Physiol. and Behav.*, **41**, 481–7.

Torrance, E.P. (1958) Sensitization versus adaption in preparation for emergencies: Prior experience with an emergency ration and its acceptability in a stimulated survival situation. *J. Appl. Psychol.*, **42**, 63.

Tuorila-Ollikainen, M., Mahlamali-Kuttanen, S., and Kurkela, R. (1984) Relative importance of color, fruity flavor, and sweetness in the overall liking of soft drinks. *J. Food Sci.*, **49**, 1598–600.

Urbanzi, G. (1982) Investigation into the interaction of different properties in the course of sensory evulation. I. The effect of colour upon the evaluation of taste in fruit and vegetable products. *Acta. Alimentaria*, **11**, 233–43.

Vickers, Z.M. (1982) Relationships of chewing sounds to judgements of crispness, crunchiness, and hardness. *J. Food Sci.*, **47**(1), 121–4.

Vickers, Z.M. (1991) Sound perceptions and food quality. *J. Food Qual.*, **14**(1), 87–96.

Wheatley, J. (1973) Putting color into marketing. *Marketing*, **67**, 26–9.

Worthington, O.J. (1960) The correlation of color measurements on canned plums with consumer acceptability. *Food Tech.*, **15**(6), 283–5.

Yoshikawa, S., Nishimaru, S., Tashiro, T., and Yoshida, M. (1970a) Collection and classification of words for description of food texture. I. Collection of words. *J. Texture Stud.*, **1**, 437–442.

Yoshikawa, S., Nishimaru, S., Tashiro, T., and Yoshida, M. (1970b) Collection and classification of words for description of food texture. II. Texture profiles. *J. Texture Stud.*, **1**, 443–51.

Yoshikawa, S., Nishimaru, S., Tashiro, T., and Yoshida, M. (1970c) Collection and classification of words for description of food texture. III. Classification of multivariate analysis. *J. Texture Stud.*, **1**, 452–63.

Zellner, D.A., Bartoli, A.M., and Eckard, R. (1991) Influence of color on odor identification and liking ratings. *Amer. J. Psych.*, **104**(4), 547–61.

Zellner, D.A., Stewart, W.F., Rozin, P., and Brown J.M. (1988) Effect of temperature and expectations on liking for beverages. *Physiol. and Behav.*, **44**, 61–8.

Zellner, D.A., Rozin, P., Aron, M., and Kulish, C. (1983) Conditioned enhancement of human's liking for flavors by pairing with sweetness. *Learning and Motiv.*, **14**, 338–50.

Index